Online Algorithms for the Portfolio Selection Problem

Robert Dochow

Online Algorithms for the Portfolio Selection Problem

With a foreword by Prof. Dr.-Ing. Günter Schmidt

Springer Gabler

Robert Dochow
Saarbrücken, Germany

Dissertation Saarland University, Saarbrücken, 2015

ISBN 978-3-658-13527-0 ISBN 978-3-658-13528-7 (eBook)
DOI 10.1007/978-3-658-13528-7

Library of Congress Control Number: 2016939257

Springer Gabler

Printed on acid-free paper

This Springer Gabler imprint is published by Springer Nature
The registered company is Springer Fachmedien Wiesbaden GmbH

To Lucie Elena

Foreword

When Robert Dochow came to the Saarland University in the year 2010 he had solid financial knowledge and a typical business administration perspective. At the chair of Operations Research and Business Informatics he acquired, through his ambitious desire for learning and understanding, a great expertise in management science and combinatorial online problems. His perspective enlarged naturally to computer science. With this he concentrated on the study of the financial portfolio selection problem. This book is one result of his efforts. It is good to see that our numerous discussions, his large experiences in teaching and supervision of students are incorporated in the development of this book.

Automated trading with algorithms is more and more evolving in financial markets, this book focusses on this subject. It contains a clearly written survey and introduction to the issues of portfolio selection, online algorithms and performance evaluation. It is competitive with the book Online Portfolio Selection: Principles and Algorithms of Li and Hoi. It is impressing that Robert Dochow designs two new algorithms for portfolio selection with mathematically provable properties. The extent and nature of the empirical investigation of algorithms for the portfolio selection problem is another value of this book. The same is true for the developed software-tool. The target readers have finance and computer science background. The book tries to help bringing people together from both communities.

I wish the book of Robert Dochow great attention in science and practice. Without any doubt it has deserved it.

Saarbrücken, December 19, 2015 Prof. Dr.-Ing. Günter Schmidt

Acknowledgments

Foremost, I would like to thank my supervisor Günter Schmidt. The perspective on economical and financial problems he taught me, is a great enrichment for me. Without him, I would never have learned this. At the same time, I am grateful that Alois Knobloch kindly agreed to be my co-supervisor. Our discussions were very helpful to make this thesis complete.

Further, I thank the group of the chair Operations Research and Business Informatics at the Saarland University in all its employment constellations - Hedi Staub, Esther Mohr, Mike Kersch, Hai-Dung Dinh, Claude Gangolf, Iftikhar Ahmad, Melanie Holländer, Pascal Schroeder and Michael Schwarz - it was a pleasure to work with you. Especially, Hedi Staub, Esther Mohr and Mike Kersch deserve a big thank you for their fruitful discussions, comments and administrative support. Many thanks to Esther Mohr, Marcel Wiedenmann and others for proofreadings.

I appreciate the feedback from students who took my courses and projects on online portfolio selection. Particularly noteworthy students on my way are Philipp-Dennis Berlitz, Thomas Dackweiler, Susanne Reiter, Nicolas Glöckner, Pascal Schroeder and Christian Leppek. I hope I have also positively contributed to their academic or professional careers and I wish them all the best.

Above all, I thank my parents and grandparents for their love and support. And not to forget: I thank my friends Sven Müller, Janine Fisselbrand, Marcel Wiedenmann, Marina Schwertner and Constanze Löh who made me feel good in the small, but nice, Saarland.

Saarbrücken, June 22, 2015 Robert Dochow

Contents

List of Abbreviations

AC	anti correlation algorithm
ALG	algorithm
APY	annual percentage yield
AR	auto-regressive model
ARMA	auto-regressive moving average model
ASTDV	annualized standard deviation
BA	best asset portfolio
BCR	best constant rebalancing portfolio
BH	buy-and-hold portfolio
BHP	buy-and-hold problem
biCP	bi-directional conversion problem
BRVBH	best reward-to-variability buy-and-hold portfolio
BRVCR	best reward-to-variability constant rebalancing portfolio
BT	bootstrap approach
BVBH	best variance buy-and-hold portfolio
BVCR	best variance constant rebalancing portfolio
cf.	confer (compare)
chap.	chapter
CP	conversion problem
CR	constant rebalancing portfolio
CRAPS	combined risk-adjusted portfolio selection algorithm
CRP	constant rebalancing problem
CSP	constraint satisfaction problem
CV	cross-validation approach
CWMR	confidence weighted mean reversion algorithm
CWMRS	confidence weighted mean reversion algorithm with standard deviation
CWMRV	confidence weighted mean reversion algorithm with variance
DDR	drawdown ratio
DJIA	Dow Jones Industrial Average dataset
e.g.	exempli gratia (for example)
EG	exponential gradient algorithm

EMH	efficient markets hypothesis
eq.	equation
et al.	et alii (and others)
etc.	et cetera (and other things)
FC	finance community
FTL	follow-the-loser
FTW	follow-the-winner
GPSP	general portfolio selection problem
i.e.	id est (that is to say)
JBT	Jarque Bera test
JK	jackknife approach
MA	moving average model
MDD	maximum drawdown
MLC	machine learning community
MRDD	maximum relative drawdown
MSCI	Morgan Stanley Capital International dataset
ND	naive diversification
NYSE	New York Stock Exchange dataset
NYSE2	New York Stock Exchange dataset 2
OLMAR	online moving average mean reversion algorithm
ONS	online Newton step algorithm
OPS	online portfolio selection
OPT	offline algorithm
OTT	one sample t-test
OZT	one sample z-test
p.	page
pp.	pages
PAMR	passive aggressive mean reversion algorithm
PAMR1	passive aggressive mean reversion algorithm type 1
PAMR2	passive aggressive mean reversion algorithm type 2
PAMR3	passive aggressive mean reversion algorithm type 3
POO	percentage of outperformance
PSP	portfolio selection problem
PST	portfolio selection tool
PTWSRT	pairwise two sample Wilcoxon signed-rank test
PTZT	pairwise two sample z-test
RAND	random portfolio
RAPS	risk-adjusted portfolio selection algorithm
RMR	robust median mean reversion algorithm
RVR	reward-to-variability ratio

RW	random walk model
s.t.	subject to
SCR	successive constant rebalanced algorithm
SCRP	semi-constant rebalancing problem
SP500	Standard & Poors 500 dataset
SPSP	semi-portfolio selection problem
SW	switching portfolio
TSX	Toronto Stock Exchange dataset
TTT	two sample t-test
TWSRT	two sample Wilcoxon signed-rank test
TZT	two sample z-test
UBH	uniform buy-and-hold portfolio
UCR	uniform constant rebalancing portfolio
uniCP	uni-directional conversion problem
UP	universal portfolio algorithm

List of Variables

α	parameter of a model or an ALG
$\bar{a}_t^{ij}$	element of $\mathbf{A}_t^{-1}$
$\bar{x}_{it}^w$	moving average price relative with window size w for asset i at the end of trading period t
$\bar{x}_t$	average market return at trading period t
β	parameter of a model or an ALG
$\breve{\rho}_{i,j}$	cross correlation between asset i and asset j
δ	parameter of an ALG
$\delta_{t\tau}^w$	Euclidean distance between price estimator with window size w and observed price at the end of trading period $(t-\tau)$
$\dot{b}_{it}$	auxiliary update portfolio proportion of asset i at trading period t
η	parameter of an ALG
γ_t^w	auxiliary variable for the calculation of RMR at trading period t
$\hat{\bar{x}}$	parameter of an ALG
$\hat{\mu}_i$	estimator of the exponential growth rate of asset i
$\hat{\tau}$	parameter of an ALG
$\hat{\mathbf{q}}_t^w$	vector of price estimators with window size w at the end of trading period t
$\hat{q}_{it}^w$	element of $\hat{\mathbf{q}}_t^w$
$\hat{x}_{it}^w$	price relative with price estimator and current price of asset i at trading period t
$\hat{\bar{x}}_t^w$	arithmetic mean of price relatives with price estimator and current price of all assets at trading period t
κ_t^w	auxiliary variable for the calculation of RMR at trading period t
λ_t	multiplier function at trading period t
χ^2	chi-squared distribution
$\mathbf{A}_t^{-1}$	inverse matrix of $\mathbf{A}_t$
$\mathbf{A}_t$	matrix with second order information for the calculation of ONS at trading period t
$\mathbf{o}_{it}$	element i of $\mathbf{o}_t$

$\mathbf{o}_t$	vector with first and second order information for the calculation of ONS at trading period t
$\mathcal{N}$	normal distribution
$\mathcal{T}$	test statistic
$\mathfrak{B}$	set of algorithms
$\mathfrak{b}$	one element of the simplex $\mathfrak{B}_m$ / portfolio allocation vector from the simplex $\mathfrak{B}_m$
$\mathfrak{b}^*$	optimal portfolio allocation from the simplex $\mathfrak{B}_m$
$\mathfrak{b}_i^\omega$	portfolio allocation of CR-expert ω for asset i
$\mathfrak{b}_i$	proportion i of the portfolio allocation vector from the simplex $\mathfrak{B}_m$
$\mathfrak{B}_m$	simplex with m decision variables
a_t	auxiliary variable for the calculation of CWMR for trading period t
b_t	auxiliary variable for the calculation of CWMR for trading period t
c_t	auxiliary variable for the calculation of CWMR for trading period t
μ	exponential growth rate of an ALG on a given market
μ_i	exponential growth rate of asset i
Ω	number of experts / algorithms
ω	counter for expert / counter for ALG
ϕ	parameter of an ALG
Ψ	finite set of real numbers
σ	standard deviation of an ALG on a given market
σ^2	variance of an ALG on a given market
σ_i	standard deviation of asset i
τ	auxiliary counter for trading period / counter for time
τ_k	toleration value for iteration k
$\mathbf{1}_t$	trading permission for time t
$\hat{\mathbf{F}}$	sample distribution
$\mathbf{b}$	sequence of portfolio allocations
$\mathbf{b}_t$	portfolio allocation at the beginning of trading period t
$\mathbf{C}$	sequence of constraints
$\mathbf{D}$	sequence of variable bounds
$\mathbf{E}_0$	threshold value
$\mathbf{P}$	constraint satisfaction problem / optimization problem
$\mathbf{Q}_\alpha$	α-quantile
$\mathbf{W}$	sequence of period wealth
$\mathbf{W}_T$	worst-case logarithmic wealth ratio
$\mathbf{X}$	sequence of decision variables / set of markets

$\mathbf{x}$	market
$\mathbf{x}_t$	market at trading period t
$\mathbf{x}_{it}$	return factor of asset i at trading period t
$\mathbf{y}_t$	input allocation for the projection at trading period t
Θ_t^i	first order information of asset i at trading period t
Θ_t^{ij}	second order information of asset i with asset j at trading period t
$\tilde{x}_t$	confidence weighted average price relative during trading period t
Ξ	finite set of real numbers
ζ_t^w	auxiliary variable for the calculation of RMR at trading period t
ζ_t	request at time t
a	minimum distance between two portfolio proportions
a_t^{ij}	element of $\mathbf{A}_t$
A_i	asset i
B^ω	ALG ω
b_{it}	portfolio proportion of asset i at trading period t and element of $\mathbf{b}_t$
C	parameter of an ALG
c	competitive ratio / performance ratio / comparative ratio of an ALG
D	difference between two observations
d	auxiliary variable
DD_t	drawdown for trading period t of an ALG on a given market
H	hypothesis system
H_0	null hypothesis
H_1	alternative hypothesis
i	counter for asset
j	counter
K	number of iterations
k	counter for iteration
m	number of assets
M_t	maximum wealth achieved by an ALG up to time t
m_t	minimum wealth achieved by an ALG up to time t
MA_{it}^w	moving average with window size w for asset i at the end of trading period t
N	number of observations / number of items in a population
POO	percentage of outperformance for a given ALG against a given benchmark
q_{it}	price of asset i at time t
Q_{it}^j	conversion rate to convert asset i into asset j at time t

R_f	annual percentage yield of a risk-free asset
RDD_t	relative drawdown for trading period t of an ALG on a given market
s_t	fraction of wealth converted from the base asset into the counter asset at time t
T	number of trading periods / investment horizon
t	counter for trading period / counter for time instant
u_i	random variable from the standard uniform distribution
U_t	auxiliary variable for the calculation of CWMRS for trading period t
v_t	sum of square deviation from the average market change at trading period t
v_{it}	variation measure of asset i at the end of trading period t
V_t	market variation at trading period t
w	window size
W_T^*	optimal terminal wealth for a given type of PSP
W_0	initial wealth
W_T	terminal wealth
W_t	period wealth at the end of trading period t
X	random variable
x	point in a set of real numbers
x^*	optimal point in a set of real numbers
x^{max}	maximum value of price x_{it}
x^{min}	minimum value of price x_{it}
x_t	holding period return of a portfolio during trading period t
X_{it}	random variable from the standard normal distribution for asset i for trading period t
Y_i^{init}	initial shares owned by an investor of asset i at time $t = 0$ before rebalancing
Y_{it}	shares owned by an investor of asset i at time t after rebalancing
y_{it}	number of shares to buy (sell) at time t
z_α	critical value with given α of a statistical model

List of Figures

List of Tables

1 Introduction

This chapter introduces terms as portfolio selection problem, algorithm and performance evaluation from the perspective of the finance community on the one side, and the machine learning community on the other side. Then, the three research questions of the thesis are motivated and formulated. The chapter concludes with an explanation of the structure of the thesis.

1.1 Preliminaries

The *portfolio selection problem* (PSP) is a decision problem in which an investor wants to invest an amount of wealth for a finite investment horizon. To solve the PSP, the investor decides how to allocate the wealth within a finite set of, at least two, assets. Each asset represents one different investment opportunity and the realized decision for an allocation is a portfolio. The total of all assets forms a market and is considered risky, if it contains at least one risky asset, what means its prices are uncertain. A change in wealth, due to the riskiness of the market, may require a new decision by the investor. A single-period PSP requires only one decision during the whole investment horizon. In contrast, a multi-period PSP requests sequentially for a decision in periodical time intervals during the investment horizon, and is therefore denoted as *online* problem.

Any decision for the PSP depends typically on the objective function of the investor. Simplified, there are three types of objective functions. (i) Return on investment: The focus is to maximize the wealth at the end of the investment horizon. (ii) Risk: The possibility of losing wealth over the investment horizon is denoted as risk. The goal is to minimize the risk. (iii) Risk-adjusted performance: The focus is a mixture of (i) and (ii). The objective is a trade-off between achievable wealth and risk. If the decision for the PSP is based on the objective (ii) or (iii), then it can be simply stated that the investor takes risk management into account.

In accordance with Borodin et al. [2004], the PSP is solvable by an algorithm with respect to one of the three types of objective functions. An algorithm for the PSP is a deterministic or randomized rule, which specifies the allocation of wealth in the market. In the literature there are two communities which differ considerably in their perspective on the PSP and thus on the resulting algorithms.

The *finance community* (FC) is mainly influenced by the seminal work of Markowitz [1952]. In line with [Elton and Gruber, 1997], the FC characterizes the PSP as follows. For one single period, a statistical model of the asset prices in the market is assumed and the optimal decision is derived, concerning the specific objective function of the investor, e.g., [Tobin, 1958], [Sharpe, 1964], [Lee, 1977], [Kraus and Litzenberger, 1976], [Vath et al., 2007] and [Garlappi et al., 2007]. The existing works with a multi-period perspective on the PSP in, for example, [Mossin, 1968], [Fama, 1970b], [Hakansson, 1970], [Hakansson, 1974], [Merton and Samuelson, 1974], [Li and Ng, 2000] and [DeMiguel et al., 2015] consider technically only a sequence of the single-period problem. The derived optimal decisions for each trading period differ only due to the different valuation of time during the full investment horizon, based on a utility function. For the input of the statistical model a forecasting model, which requires historical data to calibrate, is taken. The existing models in the literature can be differentiated according to the required historical data, e.g., asset prices [Sharpe, 1963], industry sector data [Cohen and Pogue, 1967], economic data [Chen et al., 1986], market capitalization data [Fama and French, 1992] and option prices [Conrad et al., 2013]. In short, the FC considers risk management and the objective of the investor is the optimization of a performance measure, which either quantifies the risk of the decision or its achievable wealth in relation to risk (type of objective function (ii) and (iii)). For examples, [Markowitz, 1952], [Sharpe, 1966], [Treynor, 1965], [Jensen, 1969], [Rockafellar and Uryasev, 2000] and [Lisi, 2011] provide concrete performance measures.

The *machine learning community* (MLC) is mainly grounded on the analytic work of [Kelly, 1956], [Bell and Cover, 1980], [Cover and Gluss, 1986], [Cover, 1991], [Cover and Ordentlich, 1996] and [Ordentlich and Cover, 1998]. A portfolio which is periodically rebalanced to a specific allocation (constant rebalancing portfolio) can be more beneficial than to buy a portfolio only once and hold it for the full investment horizon (buy-and-hold portfolio). Moreover, there exists an optimal allocation, which promises the largest growth of wealth in the long run, which is only known in hindsight (best constant rebalancing portfolio). Cover [1991] introduced the *concept of universality* to classify a specific type of algorithms, as provided in [Helmbold et al., 1998], [Vovk and Watkins, 1998], [Gaivoronski and Stella, 2000] and [Agarwal et al., 2006]. These algorithms provably track this best constant rebalancing portfolio on arbitrary market instances (follow-the-winner algorithms). Empirical observations, that strong rising (crashing) asset prices will drop (rise) again, motivate the MLC to develop another type of algorithms (follow-the-loser algorithms), e.g., [Borodin et al., 2004], [Li et al., 2012], [Li and Hoi, 2012], [Li et al., 2013] and [Huang et al., 2013]. According to Li and Hoi [2014], other algorithms in the MLC try to profit from repeating price patterns in the market as for example in [Györfi et al., 2007]. In general, the idea in the MLC is to develop an algorithm, which explores information in the past asset prices from the concrete market, to suggest a portfolio allocation for the next period. The algorithm itself can also consist of a statistical model combined with a forecasting model, which evaluates and calibrates itself during the investment process. Typically, the algorithms in the literature target the problem of maximizing the wealth at the end of the investment horizon (objective function (i)).

There are many ways to evaluate the performance of an algorithm for the PSP. One concept is the *statistical analysis*, which is broadly used in the FC and MLC, e.g., [Gibbons et al., 1989], [Nawrocki, 1991], [Brock et al., 1992], [DeMiguel et al., 2009] and [Schmidt et al., 2010]. The algorithm is executed on real or artificial data and its performance measures can be directly observed and analyzed with respect to statistical measures. The advantage of this concept is a fast impression of how the algorithm performs on given data. The findings can be easily compared with those of other algorithms, but the results depend always on the availability and quality of data. To avoid this dependency on data, there exists the *competitive analysis* as an alternative concept, which is widespread in the MLC, e.g., [Ordentlich and Cover, 1998], [Helmbold et al., 1998], [Agarwal et al., 2006], [Kozat and Singer, 2011] and [Mohr et al., 2014]. In line with [Blum, 1998] and [Koutsoupias and Papadimitriou, 2000], the performance of an algorithm is evaluated from a theoretical perspective in comparison to a benchmark algorithm executed by an omniscient adversary. This resulting benchmark algorithm is always optimal in hindsight and is called *offline* algorithm. An algorithm is *competitive* if a constant c can be derived, which shows that the performance of the algorithm *never* differs more than c times from the performance of the offline algorithm. The relative performance of a competitive algorithm can be guaranteed with respect to one specific performance measure and without any knowledge of statistical properties of the market.

1.2 Motivation and Research Questions

Both communities, the FC and the MLC, focus on rules that yield "optimal" decisions for the investment in risky markets. The term "optimal" in the field of portfolio selection is sometimes confusing. Clearly, in a market with only different types of risk-free assets (i.e., different time, different return on investment), an algorithm is "optimal" when it promises the maximum of wealth at the end of the investment horizon. Due to the riskiness of a market containing risky assets a focus on risk management seems to be "optimal". However, there is a gap between the FC and the MLC:

- The communication between the FC and the MLC in the literature is low to non-existent. For example, a short retrieval for citations of the seminal work of Cover [1991] reveals that contributions of typical journals of the FC, which deal with the PSP, are missing.
- Due to the concentration on the riskiness of the market, the performance evaluation is done in the FC mainly with statistical analysis. Thus, the application of the concept of competitive analysis is hard to find in the literature of this community.
- Inspired by the concept competitive analysis, the MLC concentrates on the maximization of wealth at the end of the investment horizon and avoids regular statistical assumptions. This is a possible explanation, why risk management for online portfolio selection is still an open question in the MLC, cf. [Borodin and El-Yaniv, 1998, pp. 309-311] and [Li and Hoi, 2014, p. 29].
- The algorithms considered in this thesis have not been tested systematically on empirical data in terms of risk and the trade-off between risk and return on investment. Precisely, the algorithms are mainly executed on six different datasets with little variation in the test design (For a summary of existing test designs in the considered literature see Table 6.2). Thus, more and comprehensive empirical testing is required.

This thesis contributes to the closure of the gap between the FC and the MLC. The addressees of this work are researchers and practitioners from both communities.

To bring the FC and MLC closer together, the PSP in its various forms must be explained in detail. Each form can be found in the literature as a separate problem with different notation, terms, definitions and solutions. For a better understanding, a classification of these (sub-)problems from a comprehensive perspective is provided. The problems are simplified in two ways. First, the objective of any PSP is the maximization of wealth at the end of the investment horizon. Second, any problem is solved under the assumption that the future asset prices of the market are fully known in advance (offline perspective). These two simplifications allow a clear view on the nature of the problems and a classification of them. The "quality" of the solution for any PSP must be judged by a performance evaluation. Several performance measures and test designs exist. An overview of the most frequently used ones in the FC and the MLC is given to facilitate the comprehensive evaluation of new algorithms for researchers and practitioners. Summing up, this can be formulated by the following research question:

> **Question 1: How to classify selected types of the PSP and evaluate the performance of an algorithm for the PSP?**

Li and Hoi [2014] provide an extensive survey of algorithms for the PSP from the MLC perspective. This thesis is limited to the algorithms of this survey from the type: benchmark, follow-the-winner and follow-the-loser. Critical on this survey is that the representation of some algorithms is not appropriate for a simple implementation in a trading system. A glance at the original publication of the algorithm can remedy this, but makes the comparability of their mechanisms more difficult. This motivates the author of this thesis to work on a comprehensive pseudocode representation of the algorithms to make their implementation easier for researchers and practitioners. During this work, the second research question arose. Several ways of constructing an online algorithm to solve the PSP are used. However, all algorithms in the MLC focus exclusively on the maximization of wealth at the end of the investment horizon and an associated risk management is missing. This is a complete contrast to the FC where risk management is always taken into account. This rises the idea of designing an online algorithm with an objective function that incorporates risk. Moreover, this algorithm should be competitive. The precise research question is formulated as follows:

> **Question 2: Is it possible to construct a worst-case competitive online algorithm for the PSP that takes risk management into account?**

During the study of the results for the empirical performance of online algorithms in the literature it is noticed that the algorithms from the type follow-the-loser perform extremely better in terms of terminal wealth than those of the type follow-the-winner. This in turn is surprising, since the algorithms from the type of follow-the-winner are all competitive algorithms. In addition, there are some critics on the test designs. At least the early published algorithms, all employ the same dataset. The object of research is widely the evaluation of the return on investment. In addition, through hypothesis testing it can be verified on several datasets that the excessive return on investment of the follow-the-loser algorithms cannot be based on luck. The performance measurement of risk and the evaluation of the algorithms due to their ability to generate a trade-off between risk and return on investment are addressed only occasionally. Therefore, the third research question is:

Question 3: How is the empirical performance of selected online algorithms for the PSP, in comparison to a benchmark algorithm, on a given dataset, when the performance is measured for the return on investment, risk and risk-adjusted performance?

1.3 Structure of the Thesis

The chapters of this thesis are based on each other and it is recommended to read them in numerical order.

Chapter 2 and Chapter 3 answer Question 1 and introduce several types of the PSP. In additon, selected techniques from the FC and the MLC to evaluate the performance of an algorithm that solve a PSP are surveyed. The basis for understanding all of the following chapters is given.

In Chapter 4, well known algorithms from the literature of the MLC are provided in a form which facilitate an implementation in a trading system. This contributes to the answer of Question 2 in such a way that mechanisms of existing algorithms are understood and later on used to develop a new algorithm.

Consequently, in Chapter 5 the Question 2 is solved by proposing two new competitive algorithms with risk management as a modification of the prominent *universal portfolio algorithm* (UP) of Cover [1991]. One of them has already been published by the author of this thesis with two co-authors in [Dochow et al., 2014].

Question 3 is processed by executing empirical tests. This is done in Chapter 6 and in a small section at the end of Chapter 5. The two new algorithms of Chapter 5 are also included in the numerical experiments.

Chapter 7 is an extra contribution to Question 1. There is a software tool presented as a prototype of a trading system which was built for the empirical investigation of algorithms during the work on this thesis. It should encourage researchers and practitioners to develop further software tools.

The last chapter, Chapter 8, concludes the findings of the thesis and proposes future work.

Note that due to the extensive literature on the PSP in the FC, the associated concrete problems and particular algorithms discussed in the FC are not presented in details within this thesis. Rather problems and algorithms from the MLC are in the focus.

2 Portfolio Selection Problems

The portfolio selection problem exists in a variety of forms. In this chapter the main types of the portfolio selection problem from the literature are classified and presented as mathematical programs. The types are introduced from the perspective of an investor who buys and sells shares of assets at time instants. To derive a classification structure among the considered types of the portfolio selection problem all of them are simplified in two ways. First, each type of problem aims to maximize the wealth of the investor at the end of the investment horizon. Second, the future is completely known when the specific type of portfolio selection problem is solved. Before deriving the mathematical programs, the chapter begins with the definition of essential terms. The chapter concludes with a classification structure of the considered problems based on the maximum attainable wealth of the investor.

2.1 Preliminaries

Consider a given investment horizon of an investor with $t = 0, ..., T$ time instants and $t = 1, ..., T$ trading periods. Referring to Nee and Opper [2010, p. 2106], a transaction is defined as any action that involves an exchange of an entity during the investment horizon of an investor. In financial markets these entities are units, denoted as shares which are tradeable assets, such as stocks, currencies, bonds or commodities [see Borodin et al., 2000, p. 173]. It is assumed that all shares of one asset are infinitely divisible (compare with [Copeland, 1976, p. 1150] and [Smith, 2001, p. 482]) and homogeneous such that each share of one asset is a perfect substitute of each other (see [Geltner, 1997, p. 426]). Note that a cash position is also an asset. A definition of a *transaction* is given:

Definition 2.1 *Given two assets A_1 and A_2. At time instant $(t-1)$ the investor owns $Y_{1(t-1)}$ shares of A_1 and $Y_{2(t-1)}$ of A_2. Then, at time instant t one action (transaction) is executed. It follows that $\left(Y_{1t} - Y_{1(t-1)}\right) < 0$ and $\left(Y_{2t} - Y_{2(t-1)}\right) > 0$ or vice versa.*

Consider the scenario that asset A_1 is a cash asset and A_2 is a stock asset. The investor owns at time instant $t-1$ only cash such that $Y_{1(t-1)} > 0$ and $Y_{2(t-1)} = 0$. The investor gives the entire cash to receive a finite number of A_2 shares such that $Y_{1t} = 0$ and $Y_{2t} > 0$. The purchase of shares of asset A_2 is one transaction. Now, assume the investor is selling all shares of A_2 at time instant t such that at the subsequent time instant $Y_{1(t+1)} > 0$ and $Y_{2(t+1)} = 0$. The sale of the shares of A_2 is another transaction.

Note that in practice the number of transactions doubles if A_1 and A_2 are both stock assets. The cash asset is not directly considered, such that in practice the cash asset is indirectly a unit of account to realize transactions if a direct exchange between A_1 and A_2 is technically not possible. The investor sells shares of A_1 and receives cash for it with one transaction. With the executing of one more transaction the investor buys shares of A_2. Both transactions can be realized at the same time instant t.

From this perspective, *trading* is defined as follows:

Definition 2.2 *Given m assets $A_1, ..., A_m$ with $m \geq 2$. Trading is a sequence of transactions among the m assets during the investment horizon with $t = 0, ..., T$ time instants. The sequence must include at least one transaction.*

In other words, trading is about allocating wealth. Note, if the transactions are based on specific rules and controlled by computers, then such trading is called algorithmic trading (see also Prix et al. [2007, p. 717] and Sadoghi et al. [2010, p. 1525]).

If trading is carried out by taking the optimization of a certain or at least vague objective function into account, then this problem is denoted as PSP. Note that, the term trading problem is often used as a synonym for some types of the PSP in the literature. It is assumed that the wealth of the investor is determinable at all time instants with $t = 0, ..., T$ and there is no preference of the investor for a particular asset. At any time instant, the value of wealth allocated in one asset is as good as the value of wealth allocated in another asset. Transactions at the end of the investment horizon (at time instant $t = T$) are done by the investor only for technical reasons. For example, because of the realization of profits. A change in wealth of the investor occurs if and only if the price of at least one asset is changing during time instant t and $t + 1$. Thus, a PSP is defined as:

Definition 2.3 *Given m assets $A_1, ..., A_m$ with $m \geq 2$. A PSP asks for a sequence of transactions by taking an objective function into account. The* *sequence aims a reallocation of wealth on the m assets for at least one time instant with $t = 0, ..., T$ and $T \geq 1$.*

A special case of the PSP is the two-way trading problem. El-Yaniv [1998, p. 45] defines it as a PSP with only two assets ($m = 2$) where one of the assets is a cash asset. The aim is to buy and sell shares of A_1 against shares of A_2 over time. Therefore, a two-way trading problem when ignoring the cash asset assumption is defined as follows:

Definition 2.4 *A two-way trading problem is a PSP with $m = 2$.*

In the one-way trading problem an investor asks to trade a given asset A_1 into A_2. At each time instant t a new conversion rate is announced, showing how many shares of A_2 to receive for one share of A_1. The investor must decide how many A_1 to convert into A_2 for the current conversion rate [compare with El-Yaniv, 1998; El-Yaniv et al., 2001; Zhang et al., 2012]. Thus, a one-way trading problem is defined as follows:

Definition 2.5 *A one-way trading problem is a two-way trading problem where only trading from A_1 into A_2 (or A_2 into A_1) for $t = 0, ..., T$ is allowed.*

According to Mohr et al. [2014, pp. 88, 89], the one-way trading problem and the two-way trading problem are often summarized by the term *conversion problem* (CP) and the objective is to convert all wealth from one asset into another asset or vice versa until $t = T$. Thus, a CP is defined as:

Definition 2.6 *A CP is given by a one-way trading problem or a two-way trading problem plus the condition that the entire wealth is transferred into only one target asset (A_1 or A_2) until the last time instant T.*

In the previous situation, the investor is able to execute transactions at each time instant during the investment horizon with $t = 0, ..., T$. This may not be possible, or desired by the investor. Transactions are only allowed on selected time instants, for example, because due to the individual investment policy or a limitation of transaction costs for the investor. This means in practice, for example, a weekly, monthly or yearly reallocation of wealth. With reference to Kozat and Singer [2011] such a problem is denoted as *semi-portfolio selection problem* (SPSP). A definition of the SPSP is given:

Definition 2.7 *Given a PSP with $t = 0, ..., T$ time instants where theoretically trading is possible. Let $\mathbf{1}_t \in \{0, 1\}$ showing if the specific investor is* *indeed allowed ($\mathbf{1}_t = 1$) or not allowed ($\mathbf{1}_t = 0$) to trade at time instant t. A SPSP is given if $1 \leq \sum_{t=0}^{(T-1)} \mathbf{1}_t < T$ time instants with a positive trading permission exist.*

Let a target allocation of wealth be expressed as a vector with m proportions, each quantifies the percentage of wealth which must be invested in a specific asset at one time instant. The action for the realization of an arbitrary target allocation is called rebalancing. Assume the investor searches for one unique target allocation at all time instants with $t = 0, ..., (T-1)$ that minimizes or maximizes an objective function. A target allocation at $t = T$ is not considered because it has no impact on the wealth at the end of the investment horzion. The problem where only one specific allocation must be realized at all time instants with $t = 0, ..., (T-1)$ is called *constant rebalancing problem* (CRP). For example, the algorithms of Cover [1991] and Kozat and Singer [2011] are based on such problem formulation. The CRP is defined as follows:

Definition 2.8 *Given a PSP with $t = 0, ..., T$ time instants where trading is possible. In a CRP an investor aims to realize always the same target allocation of wealth at all time instants with $t = 0, ..., (T-1)$.*

Summing up the Definitions 2.1 - 2.8, it is evident that different types of the PSP differ in the employed conditions, which address the specific problem. Such conditions are formulated as constraints of the most general PSP given in Definition 2.3. The following conditions are worked out in order to describe selected types of the PSP and to derive a classification structure of problems in the further course of the chapter:

(i) **Trading Condition**: This condition limits the number of time instants at which trading is allowed. A problem is considered as a *standard* problem when trading is allowed at each time instant with $t = 0, ..., T$. Otherwise it is considered as a *semi*-problem, cf. Definition 2.7.

(ii) **Allocation Condition**: A problem without restriction on the allocation of wealth is considered as a *standard* problem. If a condition for a target allocation for all possible time instants exists, then it is denoted as a CRP as formulated in Definition 2.8.

(iii) **Conversion Condition**: The objective of the investor is a complete sell out of one particular asset or a transfer of the entire wealth into one target asset A_i up to time instant $t = T$. This condition is motivated from Definition 2.6.

(iv) **Direction Condition**: The investor is restricted, such that a specific asset is only sellable (buyable) at all time instants with $t = 0, ..., T$. In other words, trading is allowed only in one direction. A problem without this condition is denoted as a *bi-directional* problem. However, if this restriction on the trading direction exists, then the problem is denoted as a *uni-directional* problem. The condition is motivated through Definition 2.4 and Definition 2.5 and is in line with Mohr et al. [2014].

2.1.1 Online and Offline Algorithms

A problem is a subjective intellectual challenge or situation which can be formulated as an open question [see Blum and Niss, 1991, p. 37]. A *mathematical problem* is a question, in which mathematical activities by help of mathematical objects and structures are needed to look for or build a possible solution that is not immediately accessible [refer to Godino and Batanero, 1998, p. 4]. According to Knuth [1968, pp. 1-9], a mathematical problem, which is solvable by a step-by-step procedure of calculations is called algorithmic problem. The procedure itself is called *algorithm* (ALG) when it satisfies the five characteristics:

(i) **Finiteness**: After a finite number of steps the procedure terminates.

(ii) **Definiteness**: Each step of the procedure is precisely defined such that the actions need to be carried out are specified for each possible case.

(iii) **Input**: The procedure has zero or more inputs which are taken from a specified set of mathematical objects.

(iv) **Output**: The procedure has one ore more outputs with a specified relation to the input.

(v) **Effectiveness**: All steps are basic instructions and need to be formulated clear and comprehensible. They can be done exactly and in finite length of time by someone using a pencil and paper when applying a small example.

In line with Karp [1992] and Albers [2006], algorithms are divided into *online* and *offline* algorithms:

An online ALG is a solution for an algorithmic problem where not the entire input is available from the beginning. The input is offered to the online ALG as a sequence. Each additional input causes a partial problem, generally formulated as a request. These requests must be answered by the online ALG in the order of appearance with corresponding actions. All requests and associated actions involve costs, often measured in units of time or money. The sequence of actions generated by an online ALG may be in hindsight not "optimal". More formal:

Definition 2.9 *Let $\zeta_0, ..., \zeta_T$ be a sequence of requests, which must be solved each by an action in the order of occurrence at $t = 0, ..., T$. An online ALG provides these actions as output with knowing at time instant t only the current and historical requests $\zeta_0, ..., \zeta_t$.*

In contrast, an offline ALG is an omniscient adversary, which knows the entire sequence of requests in advance. The adversary has to process each request, but the choice of each action is based not only on the historical requests, but also on the future requests. Thus, the future is ex-ante known and the sequence of actions, which minimizes the costs, is selected. Concerning the specific cost function the actions of the adversary are *optimal.* In accordance with the Definition 2.9, an offline ALG is defined as follows:

Definition 2.10 *Let $\zeta_0, ..., \zeta_T$ be a sequence of requests, which must be solved each by an action in the order of occurrence at $t = 0, ..., T$. An offline ALG provides these actions as output with knowing at time instant t (with $t < T$) the current and historical requests $\zeta_0, ..., \zeta_t$ plus all future requests $\zeta_{t+1}, ..., \zeta_T$.*

2.1.2 Mathematical Programming

Mathematical programming is a language for describing optimization problems [see Liberti et al., 2009, p. 1]. The question of an optimization problem is always to find the "best" solution from a set of feasible solutions. Such a solution is called optimal [see also Roy, 1971, p. 239].

A formal definition of an optimization problem is found in Horst and Tuy [1996, pp. 3-4] (More specifically, this is limited to a minimization problem.):

Definition 2.11 *Given a nonempty closed set $\Psi \subset \mathbb{R}^m$ and a continuous function $f(\cdot) : \Xi \to \mathbb{R}$, where $\Xi \subset \mathbb{R}^m$ is a suitable set containing Ψ. The aim is, to find at least one point $x^* \in \Psi$ satisfying $f(x^*) \leq f(x)$ for all $x \in \Psi$ or show that such a point does not exist.*

For the sake of simplicity it is often assumed that a solution $x^* \in \Psi$ exists. An optimization problem can be represented by a minimzation problem

$$\begin{aligned} &\mathbf{min}\ f(x) \\ &\text{s.t.} \quad x \in \Psi \end{aligned}$$

or a maximization problem

$$\begin{aligned} &\mathbf{max}\ f(x) \\ &\text{s.t.} \quad x \in \Psi \end{aligned}$$

where $\mathbf{max}\ f(x) = -\mathbf{min}\ (-f(x))$. The maximization problem aims to find a x^* which satisfies $f(x^*) \geq f(x)$ for all $x \in \Psi$. From this perspective it follows Remark 2.1:

Remark 2.1 *(i) Given two nonempty closed sets Ψ_1 and Ψ_2 with $\Psi_1, \Psi_2 \subset \mathbb{R}^m$ and one continuous function $f(\cdot) : \Xi \to \mathbb{R}$, where $\Xi \subset \mathbb{R}^m$ is a suitable set containing Ψ_1 and Ψ_2. Let x_1^* and x_2^* be the optimal solutions for two different optimization (minimization) problems according to Definition 2.11. Both are using the same function $f(\cdot)$. Then, if $\Psi_1 \subseteq \Psi_2$ it follows that $f(x_1^*) \geq f(x_2^*)$. (ii) Facing two optimization problems with the aim of the maximization of $f(\cdot)$. Then, if $\Psi_1 \subseteq \Psi_2$ it follows that $f(x_1^*) \leq f(x_2^*)$.*

One way to model an optimization problem is as a *mathematical program*. It allows a deeper understanding of a specific problem by a standardized perspective. The concept of a mathematical program was first illustrated by Dantzig [1948]. As mentioned in Fourer et al. [1993, p. 7], a mathematical program is easy to convert into a language that a computer can understand. And to solve the specific optimization problem only a solver and enough computing power is required. With reference to Rader [2010, pp. 1-20], in a general form it has four components:

(i) **Decision Variables**: The decision variables x represent the actions to be made to solve the problem where x^* is the optimal decision concerning the objective function.

(ii) **Objective Function**: The objective function is a function depending on the decision variables and should be minimized or maximized, i.e., the $f(\cdot)$.

(iii) **Constraints**: The constraints are a limitation of the decision variables formulated as restrictions, requirements or interactions inside the optimization problem. Consider the set Ξ containing all theoretically possible solutions. The constraints limit these solutions such that a set Ψ with the remaining solutions arises, i.e., $\Psi \subset \Xi$.

(iv) **Variable Bounds**: The variable bounds define the value range for which the decision variables have an interpretation. Like the constraints, the variable bounds reduce the number of remaining solutions in the set Ψ. In fact, the variable bounds are constraints as well.

Note that Fourer et al. [1993, pp. 6-7] consider the model parameters in the objective function and constraints as an extra component. They argue that any mathematical program is a model for a specific type of optimization problem where each combination of parameter values specifies one *instance* of that problem.

To illustrate the consequences of Remark 2.1 for a mathematical program, the Definition 2.11 must be reformulated such that it takes the above four components of [Rader, 2010, pp. 1-20] into account.

Consider a problem, containing only decision variables, constraints and variable bounds but not the objective function. Such a problem is called *constraint satisfaction problem* (CSP). According to Smith [2005, p. 1] and Tsang [1993, p. 9], a CSP is defined as follows:

Definition 2.12 *A CSP is given by a triple* $\boldsymbol{P} = \langle \boldsymbol{X}, \boldsymbol{D}, \boldsymbol{C} \rangle$ *where* $\boldsymbol{X}$ *is a sequence of decision variables* $\boldsymbol{X} = x_1, ..., x_m$ *with associated variable bounds* $\boldsymbol{D} = D_1, ..., D_m$. *And* $\boldsymbol{C} = C_1, ..., C_d$ *is a sequence of constraints. The aim is, to find at least one combination values for the decision variables* $\boldsymbol{X}'$ *that satisfy each variable bound* $\boldsymbol{D}$ *and all constraints of* $\boldsymbol{C}$.

If the CSP is extended by adding an objective function, then it occurs an optimization problem, which is describable by a mathematical program. The revised definition of an optimization problem with respect to Definition 2.12 is given on the base of Smith [2005, p. 15] and Tsang [1993, pp. 299, 300]:

Definition 2.13 *Given a constraint satisfaction problem and an objective function* $f(\cdot)$, *which maps each solution of the CSP to a numerical value, i.e., by a tuple* $\boldsymbol{P} = \langle \boldsymbol{X}, \boldsymbol{D}, \boldsymbol{C}, f(\cdot) \rangle$. *The aim is, to find at least one combination of the decision variables* $\boldsymbol{X}$, *which generates the optimal (minimal or maximal) numerical value* $f(\boldsymbol{X}^*)$.

Combine Definition 2.13 with Remark 2.1:

Definition 2.14 *Given two optimization problems* $\boldsymbol{P}1$ *and* $\boldsymbol{P}2$. *Let* $\boldsymbol{P}1 = \langle \boldsymbol{X}, \boldsymbol{D}, \boldsymbol{C}^1, f(\cdot) \rangle$ *and* $\boldsymbol{P}2 = \langle \boldsymbol{X}, \boldsymbol{D}, \boldsymbol{C}^2, f(\cdot) \rangle$ *with identical decision variables* $\boldsymbol{X}$, *variable bounds* $\boldsymbol{D}$ *and objective function* $f(\cdot)$. $\boldsymbol{X}^{1*}$ *is the optimal solution for* $\boldsymbol{P}1$ *and* $\boldsymbol{X}^{2*}$ *for* $\boldsymbol{P}2$. *If the sequence of constraints for* $\boldsymbol{P}1$ *is* $\boldsymbol{C}^1 = C_1, ..., C_k$ *and* $\boldsymbol{P}2$ *consists of one more constraint* $C_{(k+1)}$ *such that* $\boldsymbol{C}^2 = C_1, ..., C_{(k+1)}$, *then* $f(\boldsymbol{X}^{1*}) \leq f(\boldsymbol{X}^{2*})$ *if* $\boldsymbol{P}1$ *and* $\boldsymbol{P}2$ *are minimization problems or* $f(\boldsymbol{X}^{1*}) \geq f(\boldsymbol{X}^{2*})$ *if both are maximization problems.*

Take note that this definition allows to formulate a relationship between two optimization problems. It shows that one problem is a special case of another problem. Therefore, one can make the statement that the value of the objective function of one problem is always better (larger or smaller) than or equal to the value of the objective function of another problem. In other words, the general case achieves always a better or equal value for the objective function than the special case. An example is given to demonstrate the statement of Definition 2.14.

Example 2.1 *Let $f(\cdot)$ be an objective function with decision variables $\boldsymbol{X}$ and variable bounds $\boldsymbol{D}$. There are four possible constraints: C_1, C_2, C_3, C_4. One may consider the following optimization problems:*

(i) Let $\boldsymbol{P}1 = \langle \boldsymbol{X}, \boldsymbol{D}, \boldsymbol{C}^1, f(\cdot)\rangle$ be a maximization problem with $\boldsymbol{C}^1 = C_1$.

(ii) Let $\boldsymbol{P}2 = \langle \boldsymbol{X}, \boldsymbol{D}, \boldsymbol{C}^2, f(\cdot)\rangle$ be a maximization problem with $\boldsymbol{C}^2 = C_1, C_2$.

(iii) Let $\boldsymbol{P}3 = \langle \boldsymbol{X}, \boldsymbol{D}, \boldsymbol{C}^3, f(\cdot)\rangle$ be a maximization problem with $\boldsymbol{C}^3 = C_1, C_2, C_3$.

(iv) Let $\boldsymbol{P}4 = \langle \boldsymbol{X}, \boldsymbol{D}, \boldsymbol{C}^4, f(\cdot)\rangle$ be a maximization problem with $\boldsymbol{C}^4 = C_1, C_2, C_3, C_4$.

The task of Example 2.1 is to formulate a relationship between the four *abstract* optimization problems $\boldsymbol{P}1$ - $\boldsymbol{P}4$ without regarding the concrete values of the constraints C_1 - C_4. It is known that $\boldsymbol{C}^2$ consists of $\boldsymbol{C}^1$ plus one more constraint, $\boldsymbol{C}^3$ consists of $\boldsymbol{C}^2$ plus one more constraint, and so on. Thus, due to Definition 2.14, the result is: $\boldsymbol{P}2$, $\boldsymbol{P}3$ and $\boldsymbol{P}4$ are special cases of $\boldsymbol{P}1$; $\boldsymbol{P}3$ and $\boldsymbol{P}4$ are special cases of $\boldsymbol{P}2$; $\boldsymbol{P}4$ is a special case of $\boldsymbol{P}3$. The optimal values for the objective function of $\boldsymbol{P}1$ - $\boldsymbol{P}4$ are always in the following relationship: $f(\boldsymbol{X}^{1*}) \geq f(\boldsymbol{X}^{2*}) \geq f(\boldsymbol{X}^{3*}) \geq f(\boldsymbol{X}^{4*})$.

2.1.3 Asset Prices, Conversion Rates and Return Factors

The money, which is paid (received) to get (give) for one share of an asset, is the asset price. The input of all types of a PSP is basically a sequence of asset prices for $m \geq 2$ assets and at least two time instants ($T \geq 1$). But, for some types of the PSP, it seems to be beneficial to manipulate this input in such a way that conversion rates or return factors are given instead of asset prices, cf. [Ryan, 2007, pp. 523-532], [Mohr et al., 2014] and [Li and Hoi, 2014]. To clarify the terms for the input, they are regarded in more detail.

Asset Prices and Conversion Rates

Let q_{it} be the price of A_i at time instant t. Consider a market with two assets (A_i and A_j) and both are regarded as a conversion pair with A_i as base asset and A_j as counter asset (the terms are based on currency portfolios as described in [Darvas, 2009]). Assume, wealth invested in the base asset is only convertible into wealth invested in the counter asset, denoted as conversion. The conversion is quantified by the conversion rate and is calculated by

$$Q_{it}^{j} = \frac{q_{it}}{q_{jt}} \tag{2.1}$$

for time instant t. It gives the number of shares of the counter asset which are received if one share of the base asset is sold. The reciprocal of the conversion rate quantifies the shares of the base asset, which are payed to receive one share of the counter asset. In a market with $i = 1, ..., m$ assets exists $(m^2 - m)$ different conversion pairs.

An example market given by asset prices with three assets ($m = 3$) and five time instants ($T = 4$) is shown in Table 2.1. In this market, A_1 is a cash asset, A_2 is an asset with increasing prices and A_3 is one with decreasing prices. The corresponding market given by conversion rates is provided in Table 2.2.

t	0	1	2	3	4
q_{1t}	3.00	3.00	3.00	3.00	3.00
q_{2t}	1.00	2.00	3.00	4.00	5.00
q_{3t}	10.00	8.00	6.00	4.00	2.00

Table 2.1: Market given by asset prices with $m = 3$ and $T = 4$

Asset Prices and Return Factors

Alternatively, the market can be given by return factors x_{it} with $t = 1, ..., T$ trading periods and $i = 1, ..., m$ assets. Note, the time convention changes from time instants to trading periods. See for examples, Cover [1991],

t	0	1	2	3	4
Q^1_{1t}	1	1	1	1	1
Q^2_{1t}	3.000	1.500	1.000	0.750	0.600
Q^3_{1t}	0.300	0.375	0.500	0.750	1.500
Q^1_{2t}	0.333	0.667	1.000	1.333	1.667
Q^2_{2t}	1	1	1	1	1
Q^3_{2t}	0.100	0.250	0.500	1.000	2.500
Q^1_{3t}	3.333	2.667	2.000	1.333	0.667
Q^2_{3t}	10.000	4.000	2.000	1.000	0.400
Q^3_{3t}	1	1	1	1	1

Table 2.2: Market given by conversion rates with $m = 3$ and $T = 4$

Borodin et al. [2004] and Li and Hoi [2014]. A return factor is calculated by $x_{it} = \frac{q_{it}}{q_{i(t-1)}}$ and is the factor by which the price of asset A_i changes during trading period t. In Table 2.3 the market of Table 2.1 is shown as a market with return factors.

t	1	2	3	4
x_{1t}	1.00	1.00	1.00	1.00
x_{2t}	2.00	1.50	1.33	1.25
x_{3t}	0.80	0.75	0.67	0.50

Table 2.3: Market given by return factors with $m = 3$ and $T = 4$

2.2 Selected Portfolio Selection Problems

Consider a market with two homogeneous and perfectly divisible assets (A_1 and A_2) given by asset prices. The wealth of the investor is completely represented by the property of the shares of both assets. The initial wealth W_0 of the investor at time instant $t = 0$ is represented by an amount of initial shares Y_i^{init} with $i = 1, 2$. For example, suppose the given initial wealth W_0 at time instant $t = 0$ is completely invested in A_1. Then, the investor owns before rebalancing $Y_1^{init} > 0$ initial shares of A_1 and $Y_2^{init} = 0$ initial shares of A_2 at time instant $t = 0$.

The wealth at each time instant with $t = 0, ..., T$ is calculated by

$$W_t = q_{1t}Y_{1t} + q_{2t}Y_{2t} \tag{2.2}$$

where Y_{1t} and Y_{2t} is the number of shares of A_1 and A_2, owned by the investor at time instant t after rebalancing [compare to Karatzas et al., 1987, p. 1560 eq. (2.6)]. The number of shares are calculated by

$$Y_{it} = Y_{i(t-1)} + y_{it} \tag{2.3}$$

where y_{it} is the number of shares of A_i, which are traded by the execution of one transaction with $i = 1, 2$ at time instant t. A transaction contains either a buy or a sell of shares. If $y_{it} > 0$, then shares of A_i are bought, if $y_{it} < 0$, then shares of A_i are sold and if $y_{it} = 0$, no transaction for A_i is executed. The number of shares to be owned of A_i at time instant T is

$$Y_{iT} = Y_i^{init} + \sum_{t=0}^{T} y_{it} \tag{2.4}$$

and is determined by an aggregation of the recursive Equation (2.3) with the time instants $t = 0, ..., T$. Note, if $t = 0$, then $Y_{i(t-1)} = Y_i^{init}$.

The wealth at the end of the investment horizon, denoted as terminal wealth, depends on the initial shares and the shares to be traded of A_1 and A_2 during the investment horizon. It is calculated by

$$W_T = q_{1T}\left(Y_1^{init} + \sum_{t=0}^{T} y_{1t}\right) + q_{2T}\left(Y_2^{init} + \sum_{t=0}^{T} y_{2t}\right) \tag{2.5}$$

through including Equation (2.4) into Equation (2.2). The investor decides on the allocation of wealth by manipulating the variables y_{1t} and y_{2t} at each time instant $t = 0, ..., T$.

Now it is assumed that the market can consists of more than two homogeneous and perfect divisible assets, i.e., A_i with $i = 1, ..., m$. The period wealth for any allocation of wealth among the m assets at time instant t is based on the number of shares owned by the investor at time instant t, i.e., $Y_{1t}, ..., Y_{mt}$. Thus, the period wealth is calculated by

$$W_t = q_{1t}Y_{1t} + q_{2t}Y_{2t} + \cdots + q_{mt}Y_{mt} \tag{2.6}$$

at time instant t. This can be rewritten as

$$W_t = \sum_{i=1}^{m} q_{it} Y_{it} \tag{2.7}$$

and extended such that the terminal wealth is determined by

$$W_T = \sum_{i=1}^{m} q_{iT} \left(Y_i^{init} + \sum_{t=0}^{T} y_{it} \right) \tag{2.8}$$

when Equation (2.4) is included.

2.2.1 General Portfolio Selection Problem

Suppose an investor has the objective to maximize the wealth at the end of the investment horizon, as calculated in Equation (2.5). For simplification, without loss of generality it is not allowed to borrow and sell shares of A_1 and A_2 which are not owned by the investor, i.e., assumption of no short-selling. This is a limitation of the action space for the transferable wealth of the investor. It excludes the possibility of leveraging. Therefore, short-selling is prohibited and the investor owns only a positive number of shares at each time instant t, i.e., $Y_{it} \geq 0$. In fact, this constraint is necessary for online portfolio selection [compare to Li and Hoi, 2014, pp. 4-7] and nevertheless realistic for individual investors [see Jones and Lamont, 2002]. Thus, by rearranging Equation (2.4), short-selling is excluded if

$$Y_i^{init} + \sum_{t=0}^{\tau} y_{it} \geq 0 \tag{2.9}$$

is true for all $i = 1, 2$ and time instants with $\tau = 0, ..., T$.

In addition, assume the investor does not take out or add wealth during the whole investment horizon ($t = 0, ..., T$), that is the no consumption assumption [see "self-financed" in Li and Hoi, 2014, pp. 4-7]. The wealth, taking out from A_1 is directly converted into wealth invested in A_2 and vice versa. The number of shares to be bought (sold) of A_1 are directly converted into a number of shares to be sold (bought) of A_2. Therefore, the equation

$$y_{2t} = -y_{1t} \frac{q_{1t}}{q_{2t}} \tag{2.10}$$

ensures that consumption is excluded. Note that the number of shares to be traded between A_1 and A_2 are transformed by the conversion rate, i.e.,

$\frac{q_{1t}}{q_{2t}} = Q^2_{1t}$. In the absence of transaction costs, a conversion itself does not change the current wealth. The wealth is only effected if the asset prices are changing between time instant t and $t+1$. Thus, a decision in $t = T$ does not have an effect on the terminal wealth W_T.

By including Equation (2.10) into Equation (2.5) the terminal wealth is calculated by

$$W_T = q_{1T}\left(Y_1^{init} + \sum_{t=0}^{T} y_{1t}\right) + q_{2T}\left(Y_2^{init} + \sum_{t=0}^{T} -y_{1t}\frac{q_{1t}}{q_{2t}}\right) \tag{2.11}$$

and depends only on the one sequence with $y_{11}, ..., y_{1T}$.

Taking into account Equation (2.11) as objective function of the investor plus the assumptions of no short-selling and no consumption, then the PSP in its most general form is defined and is denoted as *general portfolio selection problem* (GPSP). The GPSP with $m = 2$ based on a market given by asset prices for A_1 and A_2 is shown in Mathematical Program 1. No short-selling is fulfilled by rearranging Inequality (2.9) and adding it as two constraints, one for each asset.

Mathematical Program 1 - GPSP with $m = 2$ assets

Given:	$q_{i0}, ..., q_{iT}$	$, i = 1, 2$	asset prices
	Y_i^{init}	$, i = 1, 2$	shares to be owned of A_i at $t = 0$
Find:	$y_{10}, ..., y_{1T}$		shares to trade of A_1 and A_2

max $W_T = q_{1T}\left(Y_1^{init} + \sum_{t=0}^{T} y_{1t}\right) + q_{2T}\left(Y_2^{init} + \sum_{t=0}^{T} -y_{1t}\frac{q_{1t}}{q_{2t}}\right)$

s. t.

(I) $\sum_{t=0}^{\tau} y_{1t} \geq -Y_1^{init} \quad , \tau = 0, ..., T$

(II) $\sum_{t=0}^{\tau} -y_{1t}\frac{q_{1t}}{q_{2t}} \geq -Y_2^{init} \quad , \tau = 0, ..., T$

(III) $y_{1t} \in \mathbb{R} \quad , t = 0, ..., T$

To illustrate the application of Mathematical Program 1, it follows an example with two assets:

Example 2.2 *Let there be a market given by asset prices with two assets ($m = 2$) and five time instants ($T = 4$) (see Table 2.4). Asset A_1 is a cash asset with no change in price. The price of A_2 is increasing or decreasing with different degrees at each time instant t. The investor owns two shares of A_1 and none of A_2 at time instant $t = 0$, i.e., $Y_1^{init} = 2$ and $Y_2^{init} = 0$. (Note that all following calculations based on this example can be solved with Microsoft Excel including the Standard Excel Solver.)*

t	0	1	2	3	4
q_{1t}	50.00	50.00	50.00	50.00	50.00
q_{2t}	10.00	20.00	9.00	30.00	11.00

Table 2.4: Market of Example 2.2

Table 2.5 shows the optimal solution for Mathematical Program 1 using Example 2.2. The terminal wealth W_T of the investor is maximized. The investor needs to allocate at each time instants t for all $t < T$ the full amount of wealth to the asset with the largest increase in price between time instant t and $t+1$. This results into the optimal terminal wealth $W_T = 666.67$ where y_{1T} can be chosen arbitrary. Note that any extension of Mathematical Program 1 by adding more constraints will always lead into an equal or lower value for W_T than 666.67 on that given market. Therefore, the solution of Mathematical Program 1 is an upper bound for all thinkable types of a PSP with $m = 2$ assets.

t	0	1	2	3	$T = 4$
q_{1t}	50.00	50.00	50.00	50.00	50.00
q_{2t}	10.00	20.00	9.00	30.00	11.00
Q_{1t}^2	5.00	2.50	5.56	1.67	4.55
Q_{2t}^1	0.20	0.40	0.18	0.60	0.22
y_{1t}	-2.0000	4.0000	-4.0000	13.3333	0.0000
y_{2t}	10.0000	-10.0000	22.2222	-22.2222	0.0000
Y_{1t}	0.0000	4.0000	0.0000	13.3333	13.3333
Y_{2t}	10.0000	0.0000	22.2222	0.0000	0.0000
W_t	100.00	200.00	200.00	666.67	666.67

Table 2.5: Optimal solution for the GPSP using Example 2.2

As in the case with two assets, in the GPSP with m assets the investor aims to maximize the terminal wealth as calculated in Equation (2.8). Short-selling is excluded if Equation (2.9) is true for all $i = 1, ..., m$ at all time instants with $\tau = 0, ..., T$. In addition, consumption must be prohibited during all time instants with $t = 0, ..., T$. Therefore, if wealth of A_j calculated by $y_{jt}q_{jt}$ is transferred into wealth of A_i calculated by $y_{it}q_{it}$, then the sum has to be $y_{jt}q_{jt} + y_{it}q_{it} = 0$. The generalization for m assets is

$$\sum_{i=1}^{m} y_{it}q_{it} = 0 \tag{2.12}$$

with $t = 0, ..., T$.

Mathematical Program 2 describes the GPSP with m assets:

Mathematical Program 2 - GPSP with m assets

Given:	$q_{i0}, ..., q_{iT}$	$, i = 1, ..., m$	asset prices
	Y_i^{init}	$, i = 1, ..., m$	shares to be owned of A_i at $t = 0$
Find:	$y_{i0}, ..., y_{iT}$	$, i = 1, ..., m$	shares to trade of A_i

max $W_T = \sum_{i=1}^{m} q_{iT} \left(Y_i^{init} + \sum_{t=0}^{T} y_{it}\right)$

s. t.

(I) $\sum_{t=0}^{\tau} y_{it} \geq -Y_i^{init}$ $, \tau = 0, ..., T, i = 1, ..., m$

(II) $\sum_{i=1}^{m} y_{it}q_{it} = 0$ $, t = 0, ..., T$

(III) $y_{it} \in \mathbb{R}$ $, t = 0, ..., T, i = 1, ..., m$

To illustrate the application of Mathematical Program 2, it follows an example with three assets:

Example 2.3 *Let there be a market given by asset prices with three assets ($m = 3$) and five time instants ($T = 4$) (see Table 2.6). Asset A_1 is a cash asset with no change in price. The price of A_2 and A_3 are increasing or decreasing with different degrees at each time instant t and for each asset. The prices of A_2 and A_3 are negative correlated. This means, when the price of A_2 goes up (down), then the price of A_3 goes down (up). The investor owns two initial shares of A_1 and none of A_2 and A_3 at time instant $t = 0$, i.e., $Y_1^{init} = 2$, $Y_2^{init} = 0$ and $Y_3^{init} = 0$. (Note that all following calculations based on this example can be solved with Microsoft Excel including the Standard Excel Solver.)*

t	0	1	2	3	4
q_{1t}	50.00	50.00	50.00	50.00	50.00
q_{2t}	10.00	20.00	9.00	30.00	11.00
q_{3t}	10.00	5.00	9.00	4.00	8.00

Table 2.6: Market of Example 2.3

The optimal solution of Mathematical Program 2 using Example 2.3 is given in Table 2.7. The scenario is as follows: The investor starts at $t = 0$ with an initial allocation of $Y_1^{init} = 2$, $Y_2^{init} = 0$ and $Y_3^{init} = 0$. At $t = 0$ the investor is selling all shares of the cash asset A_1. Basically the investor shifts all wealth at time instant $t = 0, ..., (T-1)$ to the asset with the highest increase in price between time instant t and $t+1$. This is A_2 in $t = 0$, A_3 in $t = 1$, A_2 in $t = 2$ and A_3 in $t = 3$. The investor achieves a terminal wealth of $W_T = 2400.00$. Further, the wealth is always transferred by the principle of all-or-nothing, as it can be seen in the current shares Y_{1t}, Y_{2t} and Y_{3t}.

t	0	1	2	3	$T = 4$
q_{1t}	50.00	50.00	50.00	50.00	50.00
q_{2t}	10.00	20.00	9.00	30.00	11.00
q_{3t}	10.00	5.00	9.00	4.00	8.00
y_{1t}	-2.0000	0.0000	0.0000	0.0000	0.0000
y_{2t}	10.0000	-10.0000	40.0000	-40.0000	0.0000
y_{3t}	0.0000	40.0000	-40.0000	300.0000	0.0000
Y_{1t}	0.0000	0.0000	0.0000	0.0000	0.0000
Y_{2t}	10.0000	0.0000	40.0000	0.0000	0.0000
Y_{3t}	0.0000	40.0000	0.0000	300.0000	300.0000
W_t	100.00	200.00	360.00	1200.00	2400.00

Table 2.7: Optimal solution for the GPSP using Example 2.3

2.2.2 Constant Rebalancing Problem

When constant rebalancing is done, the current allocation of wealth is rebalanced to a specific target allocation at each time instant with $t = 0, ..., (T-1)$. An allocation of wealth is a set of m proportions of wealth, which quantifies the invested wealth in each asset in relation to the period wealth for one specific time instant. The proportion of wealth invested in A_i at all time instants with $t = 0, ..., (T-1)$ (at the beginning of each trading period with $t = 1, ..., T$) is defined as

$$b_{it} = \frac{Y_{i(t-1)} q_{i(t-1)}}{W_{(t-1)}} \tag{2.13}$$

whereby always $\sum_{i=1}^{m} b_{it} = 1$ is true, because $\sum_{i=1}^{m} Y_{i(t-1)} q_{i(t-1)} = W_{(t-1)}$ (see Equation (2.6)). Note that for technical reasons the proportion b_{it} is always realized at time instant $t-1$, i.e., beginning of trading period t. Using Equation (2.13) and combining it with Equation (2.4) the proportion of A_i at time instant τ is

$$b_{i\tau} = \frac{\left(Y_i^{init} + \sum_{t=0}^{(\tau-1)} y_{it}\right) q_{i(\tau-1)}}{W_{(\tau-1)}} \tag{2.14}$$

with $\tau = 1, ..., T$. Constant rebalancing is fulfilled if $b_{i(t+1)} = b_{it}$ at all time instants with $t = 0, ..., (T-1)$ and $i = 1, ..., (m-1)$. Thus,

$$\frac{\left(Y_i^{init} + \sum_{t=1}^{\tau} y_{it}\right) q_{i\tau}}{W_\tau} = \frac{\left(Y_i^{init} + \sum_{t=0}^{(\tau-1)} y_{it}\right) q_{i(\tau-1)}}{W_{(\tau-1)}}. \tag{2.15}$$

With reference to [Li and Hoi, 2014, pp. 9], if the objective is to maximize the terminal wealth by manipulating the variables y_{it} such that $b_{i(t+1)} = b_{it}$ is always fulfilled, then this is called *constant rebalancing problem* (CRP). Mathematical Program 3 describes the CRP with m assets when the market is given by asset prices. As shown in Cover [1991], there exists one allocation $b_{1t}, ..., b_{mt}$ with $t = 1, ..., T$ where the terminal wealth W_T is maximized.

Mathematical Program 3 - CRP with m assets

Given: $q_{i0}, ..., q_{iT}$, $i = 1, ..., m$ asset prices
Y_i^{init} , $i = 1, ..., m$ shares to be owned of A_i at $t = 0$

Find: $y_{i0}, ..., y_{iT}$, $i = 1, ..., m$ shares to trade of A_i

max $W_T = \sum_{i=1}^{m} q_{iT} \left(Y_i^{init} + \sum_{t=0}^{T} y_{it}\right)$

s. t.

(I) $\sum_{t=0}^{\tau} y_{it} \geq -Y_i^{init}$, $\tau = 0, ..., T$, $i = 1, ..., m$

(II) $\sum_{i=1}^{m} y_{it} q_{it} = 0$, $t = 0, ..., T$

(III) $\frac{\left(Y_i^{init} + \sum_{t=0}^{\tau} y_{it}\right) q_{i\tau}}{W_\tau} = \frac{\left(Y_i^{init} + \sum_{t=0}^{(\tau-1)} y_{it}\right) q_{i(\tau-1)}}{W_{(\tau-1)}}$
with $W_\tau = \sum_{i=1}^{m} \left(q_{i\tau} Y_i^{init} + q_{i\tau} \sum_{t=0}^{\tau} y_{it}\right)$
, $\tau = 1, ..., T$, $i = 1, ..., (m-1)$

(IV) $y_{it} \in \mathbb{R}$, $t = 0, ..., T$, $i = 1, ..., m$

An example for the application of Mathematical Program 3 is given in Table 2.8 when using Example 2.2. To maximize the terminal wealth W_T

the investor needs to allocate y_{1t} (and y_{2t}) for $t = 0, ..., (T-1)$ such that $b_{1(t+1)} = b_{1t}$. This results into the optimal terminal wealth $W_T = 161.10$ with $b_{1t} = 0.48$ and $b_{2t} = 0.52$ for trading period $t = 1, ..., T$.

t	0	1	2	3	$T-1$
q_{1t}	50.00	50.00	50.00	50.00	50.00
q_{2t}	10.00	20.00	9.00	30.00	11.00
Q_{1t}^2	5.00	2.50	5.56	1.67	4.55
Q_{2t}^1	0.20	0.40	0.18	0.60	0.22
y_{1t}	-1.0378	0.4993	-0.4171	1.2645	-0.7588
y_{2t}	5.1891	-1.2482	2.3173	-2.1075	3.4491
Y_{1t}	0.9622	1.4615	1.0444	2.3089	1.5501
Y_{2t}	5.1891	3.9409	6.2582	4.1506	7.5997
W_t	100.00	151.89	108.54	239.96	161.10
b_{1t}	-	0.48	0.48	0.48	0.48
b_{2t}	-	0.52	0.52	0.52	0.52

Table 2.8: Optimal solution for the CRP using Example 2.2

In Table 2.9 an example for the application of Mathematical Program 3 using Example 2.3 is given. The largest terminal wealth of $W_T = 314.81$ is realized with $b_{2t} = 0.52$ in A_2 and $b_{3t} = 0.48$ in A_3 at the beginning of each trading period with $t = 1, ..., T$. In the optimal solution there is no wealth remaining in the cash asset A_1, i.e., $b_{1t} = 0$ at the beginning of trading period $t = 1, ..., T$.

2.2.3 Semi-Portfolio Selection Problem

Assume the investor is allowed to rebalance the portfolio only at specific time instants during the investment horizon $t = 0, ..., T$ as described for example in [Kozat and Singer, 2011] and [Li and Hoi, 2014, pp. 9]. At the remaining time instants the investor is prohibited to trade any of the assets. Let be $\mathbf{1}_t = \mathbf{1}_0, ..., \mathbf{1}_T$ with $\mathbf{1}_t \in \{0, 1\}$ denote the trading permission at time instant t. If $\mathbf{1}_t = 1$, the investor is allowed to buy and sell shares of A_i such that $y_{it} \in \mathbb{R}$. However, if $\mathbf{1}_t = 0$, then trading is excluded by the condition $y_{it} = 0$. Trading is possible if $\sum_{t=0}^{T-1} \mathbf{1}_t \geq 1$ is true. This *semi-portfolio selection problem* (SPSP) is equal to the GPSP when at each

t	0	1	2	3	$T=4$
q_{1t}	50.00	50.00	50.00	50.00	50.00
q_{2t}	10.00	20.00	9.00	30.00	11.00
q_{3t}	10.00	5.00	9.00	4.00	8.00
y_{1t}	-2.0000	0.0000	0.0000	0.0000	0.0000
y_{2t}	5.2097	-1.8717	4.7970	-3.3773	10.1522
y_{3t}	4.7903	7.4868	-4.7970	25.3299	-13.9593
Y_{1t}	0.0000	0.0000	0.0000	0.0000	0.0000
Y_{2t}	5.2097	3.3380	8.1350	4.7577	14.9099
Y_{3t}	4.7903	12.2771	7.4801	32.8100	18.8507
W_t	100.00	128.15	140.54	273.97	314.81
b_{1t}	-	0.00	0.00	0.00	0.00
b_{2t}	-	0.52	0.52	0.52	0.52
b_{3t}	-	0.48	0.48	0.48	0.48

Table 2.9: Optimal solution for the CRP using Example 2.3

time instants t trading is allowed, such that $\sum_{t=0}^{T} \mathbf{1}_t = T+1$. Note that any SPSP can be transformed into a GPSP by reducing the number of trading periods through the exclusion of all time instants where trading is not allowed. Mathematical Program 4 describes the SPSP when a market is given by asset prices:

Mathematical Program 4 - SPSP with m assets

Given: $q_{i0}, ..., q_{iT}$, $i = 1, ..., m$ asset prices
Y_i^{init} , $i = 1, ..., m$ shares to be owned of A_i at $t = 0$
$\mathbf{1}_0, ..., \mathbf{1}_T$ time instant where trading is allowed ($\mathbf{1}_t = 1$) and excluded ($\mathbf{1}_t = 0$) with $\sum_{t=0}^{T-1} \mathbf{1}_t \geq 1$

Find: $y_{i0}, ..., y_{iT}$, $i = 1, ..., m$ shares to trade of A_i

max $W_T = \sum_{i=1}^{m} q_{iT} \left(Y_i^{init} + \sum_{t=0}^{T} y_{it} \right)$

s. t.
(I) $\sum_{t=0}^{\tau} y_{it} \geq -Y_i^{init}$, $\tau = 0, ..., T$, $i = 1, ..., m$
(II) $\sum_{i=1}^{m} y_{it} q_{it} = 0$, $t = 0, ..., T$
(III) $y_{it} = 0$, $t = 0, ..., T$, $i = 1, ..., m : \mathbf{1}_t = 0$
(IV) $y_{it} \in \mathbb{R}$, $t = 0, ..., T$, $i = 1, ..., m$

Table 2.10 shows the application of Mathematical Program 4 with $m = 2$ where trading is allowed at $t = 0, 1, 2, 4$ but not at $t = 3$ using Example 2.2. The terminal wealth W_T is 244.44 and is much lower than the W_T of Mathematical Program 1 ($W_T = 666.67$). This is because the risky asset A_2 can not be sold in $t = 3$ to buy shares of the cash asset A_1. Table 2.11 illustrates the application of Mathematical Program 4 on Example 2.3.

t	0	1	2	3	$T = 4$
q_{1t}	50.00	50.00	50.00	50.00	50.00
q_{2t}	10.00	20.00	9.00	30.00	11.00
Q^2_{1t}	5.00	2.50	5.56	1.67	4.55
Q^1_{2t}	0.20	0.40	0.18	0.60	0.22
$\mathbf{1}_t$	1	1	1	0	1
y_{1t}	-2.0000	4.0000	-4.0000	0.0000	0.0000
y_{2t}	10.0000	-10.0000	22.2222	0.0000	0.0000
Y_{1t}	0.0000	4.0000	0.0000	0.0000	0.0000
Y_{2t}	10.0000	0.0000	22.2222	22.2222	22.2222
W_t	100.00	200.00	200.00	666.67	244.44
b_{1t}	-	0.00	1.00	0.00	0.00
b_{2t}	-	1.00	0.00	1.00	1.00

Table 2.10: Optimal solution for the SPSP using Example 2.2

2.2.4 Semi-Constant Rebalancing Problem

Assume the investor is not allowed to trade at any time instants with $t = 0, ..., T$, but aims to maximize the terminal wealth W_T by using a target allocation of wealth. At the time instant when the investor is not allowed to trade, it is not possible to reallocate the asset proportions such that the portfolio is rebalanced to the specific target allocation. The problem is described by Kozat and Singer [2011] and defined as *semi-constant rebalancing problem* (SCRP). More formal: Trading is not executed in some time instants such that b_{it}, the proportion for A_i at time instant $(t-1)$, is not reallocated to a specific target proportion. Therefore, if the prices of the assets change and $\sum_{t=0}^{(T-1)} \mathbf{1}_t < T$, then it may happen with $0 < b_{it} < 1$ that $b_{i(t+1)} = b_{it}$ and for all $t = 1, ..., T$ with $i = 1, ..., m$ can not be true. In other words, in at least one period, it is not allowed to trade such that rebalancing to a target

t	0	1	2	3	$T = 4$
q_{1t}	50.00	50.00	50.00	50.00	50.00
q_{2t}	10.00	20.00	9.00	30.00	11.00
q_{3t}	10.00	5.00	9.00	4.00	8.00
$\mathbf{1}_t$	1	1	1	0	1
y_{1t}	-2.0000	0.0000	0.0000	0.0000	0.0000
y_{2t}	10.0000	-10.0000	40.0000	0.0000	0.0000
y_{3t}	0.0000	40.0000	-40.0000	0.0000	0.0000
Y_{1t}	0.0000	0.0000	0.0000	0.0000	0.0000
Y_{2t}	10.0000	0.0000	40.0000	40.0000	40.0000
Y_{3t}	0.0000	40.0000	0.0000	0.0000	0.0000
W_t	100.00	200.00	360.00	1200.00	440.00
b_{1t}	-	0.00	0.00	0.00	0.00
b_{2t}	-	1.00	0.00	1.00	1.00
b_{3t}	-	0.00	1.00	0.00	0.00

Table 2.11: Optimal solution for the SPSP using Example 2.3

allocation is not possible. To solve this, the equality of the proportions (as described in Equation (2.15) for the CRP) must be restricted to the time instants t where $\mathbf{1}_t = 1$, and ignored for the time instants t where $\mathbf{1}_t = 0$. For the reason of simplification trading at time instant $t = 0$ is always allowed, i.e., $\mathbf{1}_0 = 1$. Let $\dot{b}_{1t}$ be an auxiliary update proportion, which is required for technical reasons. It calculates the proportion by Equation (2.13) when $\mathbf{1}_{(t-1)} = 1$ and uses the update proportion of the previous trading period $\dot{b}_{1(t-1)}$ when $\mathbf{1}_{(t-1)} = 0$. The auxiliary update proportion for A_i for time instant $t-1$ (at the beginning of trading period t) is formulated as

$$\dot{b}_{it} = \mathbf{1}_{(t-1)} b_{it} + (1 - \mathbf{1}_{(t-1)}) \dot{b}_{i(t-1)} \tag{2.16}$$

such that the rebalancing restriction of Equation (2.13) can be reformulated to $\dot{b}_{1(t+1)} = \dot{b}_{1t}$. Mathematical Program 5 uses this update proportion (Equation (2.16)) and represents the SCRP with m assets.

Mathematical Program 5 - SCRP with m assets

Given:	$q_{i0}, ..., q_{iT}$	$, i = 1, ..., m$	asset prices
	Y_i^{init}	$, i = 1, ..., m$	shares to be owned of A_i at $t = 0$
	$\mathbf{1}_0, ..., \mathbf{1}_T$		time instant where trading is allowed ($\mathbf{1}_t = 1$) and excluded ($\mathbf{1}_t = 0$) with $\sum_{t=0}^{T-1} \mathbf{1}_t \geq 1$
Find:	$y_{i0}, ..., y_{iT}$	$, i = 1, ..., m$	shares to trade of A_i

$$\max \quad W_T = \sum_{i=1}^{m} q_{iT} \left(Y_i^{init} + \sum_{t=0}^{T} y_{it} \right)$$

s. t.

(I) $\sum_{t=0}^{\tau} y_{it} \geq -Y_i^{init}$ $\quad , \tau = 0, ..., T, \; i = 1, ..., m$

(II) $\sum_{i=1}^{m} y_{it} q_{it} = 0$ $\quad , t = 0, ..., T$

(III) $\dot{b}_{i(\tau+1)} = \dot{b}_{i\tau}$ $\quad , \tau = 1, ..., (T-1), \; i = 1, ..., (m-1)$

with $\dot{b}_{i\tau} = \mathbf{1}_{(\tau-1)} b_{i\tau} + (1 - \mathbf{1}_{(\tau-1)}) \dot{b}_{i(\tau-1)}$

and $\dot{b}_{i0} = 0$

and $b_{i\tau} = \frac{\left(Y_i^{init} + \sum_{t=0}^{(\tau-1)} y_{it} \right) q_{i(\tau-1)}}{W_{(\tau-1)}}$

and $W_\tau = \sum_{i=1}^{m} q_{i\tau} \left(Y_i^{init} + \sum_{t=0}^{\tau} y_{it} \right)$

(IV) $y_{it} = 0$ $\quad , t = 0, ..., T, \; i = 1, ..., m : \mathbf{1}_t = 0$

(V) $y_{it} \in \mathbb{R}$ $\quad , t = 0, ..., T, \; i = 1, ..., m$

Table 2.12 gives the optimal solution for the Example 2.2 using Mathematical Program 5 with $m = 2$ assets where trading is allowed at time instant $t = 0, 1, 2, 4$ but not at $t = 3$. The terminal wealth W_T is 121.49 and is lower than the W_T for the CRP showing in Table 2.8 ($W_T = 161.10$). The constant proportion for A_1 is with $\dot{b}_{1t} = 0.40$ lower than for the CRP ($b_{1t} = 0.48$). For the sake of completeness b_{it} and $\dot{b}_{it}$ with $i = 1, 2$ are also illustrated. In $t = 3$ it can be seen that b_{1t} and $\dot{b}_{1t}$ essentially differ. Therefore, excluding one trading period has already a large effect on the optimal solution. In Table 2.13 the application of the SCRP using Mathematical Program 5 on Example 2.3 is given. Comparing the result with this of the CRP showing in Table 2.9, it is noticeable that the terminal wealth reduces to $W_T = 150.92$ (for CRP it is $W_T = 314.81$). In addition, the optimal proportion changes such that the proportion of A_2 increases at the expense of A_3.

t	0	1	2	3	$T = 4$
q_{1t}	50.00	50.00	50.00	50.00	50.00
q_{2t}	10.00	20.00	9.00	30.00	11.00
Q_{1t}^2	5.00	2.50	5.56	1.67	4.55
Q_{2t}^1	0.20	0.40	0.18	0.60	0.22
$\mathbf{1}_t$	1	1	1	0	1
y_{1t}	-1.2003	0.4799	-0.4224	0.0000	0.1143
y_{2t}	6.0016	-1.1998	2.3466	0.0000	-0.5197
Y_{1t}	0.7997	1.2796	0.8572	0.8572	0.9716
Y_{2t}	6.0016	4.8018	7.1484	7.1484	6.6287
W_t	100.00	160.02	107.20	257.31	121.49
$\dot{b}_{1t}$	0.00	0.40	0.40	0.40	0.40
$\dot{b}_{2t}$	0.00	0.60	0.60	0.60	0.60
b_{1t}	-	0.40	0.40	0.17	0.40
b_{2t}	-	0.60	0.60	0.83	0.60

Table 2.12: Optimal solution for the SCRP using Example 2.2

2.2.5 Buy-and-Hold Problem

Assume the investor decides only once at time instant $t = 0$ about the number of shares to own of the assets $A_1, ..., A_m$. At the time instants $t = 1, ..., (T-1)$, the investor does not reallocate the wealth among the assets such that no buy or sell of shares is executed. At the end of the investment horizon (at time instant $t = T$) the investor is allowed to reallocate the wealth but without any effect on the terminal wealth, when transaction costs are ignored. This problem is called *buy-and-hold problem* (BHP) [see Li and Hoi, 2014, p. 7]. The investor decides only for the variables y_{i0} and y_{iT} with $i = 1, ..., m$, i.e., at time instant $t = 0$ and $t = T$. For all y_{it} with time instant $t = 1, ..., (T-1)$ there is no trade executed, i.e., $y_{it} = 0$. Therefore, the BHP can be formulated as a SPSP with $\mathbf{1}_0 = 1$, $\mathbf{1}_T = 1$ and $\mathbf{1}_t = 0$ for $t = 1, ..., (T-1)$.

Mathematical Program 6 represents the BHP with m assets.

t	0	1	2	3	$T=4$
q_{1t}	50.00	50.00	50.00	50.00	50.00
q_{2t}	10.00	20.00	9.00	30.00	11.00
q_{3t}	10.00	5.00	9.00	4.00	8.00
$\mathbf{1}_t$	1	1	1	0	1
y_{1t}	-2.0000	0.0000	0.0000	0.0000	0.0000
y_{2t}	6.0436	-1.7933	5.0448	0.0000	-1.0628
y_{3t}	3.9564	7.1733	-5.0448	0.0000	1.4614
Y_{1t}	0.0000	0.0000	0.0000	0.0000	0.0000
Y_{2t}	6.0436	4.2503	9.2951	9.2951	8.2323
Y_{3t}	3.9564	11.1296	6.0848	6.0848	7.5462
W_t	100.00	140.65	138.42	303.19	150.92
$\dot{b}_{1t}$	0.00	0.00	0.00	0.00	0.00
$\dot{b}_{2t}$	0.00	0.60	0.60	0.60	0.60
$\dot{b}_{3t}$	0.00	0.40	0.40	0.40	0.40
b_{1t}	-	0.00	0.00	0.00	0.00
b_{2t}	-	0.60	0.60	0.92	0.60
b_{3t}	-	0.40	0.40	0.08	0.40

Table 2.13: Optimal solution for the SCRP using Example 2.3

Mathematical Program 6 - BHP with m assets

Given: $q_{i0}, ..., q_{iT}$, $i = 1, ..., m$ asset prices
Y_i^{init} , $i = 1, ..., m$ shares to be owned of A_i in $t = 0$

Find: y_{i0}, y_{iT} , $i = 1, ..., m$ shares to trade of A_i in $t = 0, T$

$$\textbf{max} \quad W_T = \sum_{i=1}^{m} q_{iT} \left(Y_i^{init} + \sum_{t=0}^{T} y_{it} \right)$$

$$\begin{array}{llll}
\textbf{s. t.} & \text{(I)} & \sum_{t=0}^{\tau} y_{it} \geq -Y_i^{init} & , \tau = 0, ..., T,\ i = 1, ..., m \\
& \text{(II)} & \sum_{i=1}^{m} y_{it} q_{it} = 0 & , t = 0, ..., T \\
& \text{(III)} & y_{it} = 0 & , t = 1, ..., (T-1),\ i = 1, ..., m \\
& \text{(IV)} & y_{it} \in \mathbb{R} & , t = 0, T,\ i = 1, ..., m
\end{array}$$

An example for the application of Mathematical Program 6 is given in Table 2.14 when using Example 2.2. It can be seen that the investor holds only one asset for the full investment horizon. This is the one with the largest value for $\frac{q_{iT}}{q_{i0}}$ with $i = 1, 2$, which is the best asset in the market during the investment horizon. Obviously, the search for the best buy-and-hold portfolio when maximizing terminal wealth is always equal to the search for the best asset. Note that the best *buy-and-hold portfolio* (BH) contains only one asset if the objective is to maximize terminal wealth, such that $b_{it} \in \{0,1\}$. In this case, $b_{1(t+1)} = b_{1t}$ for all $t = 1, ..., (T-1)$ is always true. It is clear that the best asset is also part of the solution space of the CRP. Table 2.15 shows the application of the BHP with Example 2.3.

t	0	1	2	3	$T=4$
q_{1t}	50.00	50.00	50.00	50.00	50.00
q_{2t}	10.00	20.00	9.00	30.00	11.00
Q^2_{1t}	5.00	2.50	5.56	1.67	4.55
Q^1_{2t}	0.20	0.40	0.18	0.60	0.22
$\mathbf{1}_t$	1	0	0	0	1
y_{1t}	-2.0000	0.0000	0.0000	0.0000	0.0000
y_{2t}	10.0000	0.0000	0.0000	0.0000	0.0000
Y_{1t}	0.0000	0.0000	0.0000	0.0000	0.0000
Y_{2t}	10.0000	10.0000	10.0000	10.0000	10.0000
W_t	100.00	200.00	90.00	300.00	110.00
b_{1t}	-	0.00	0.00	0.00	0.00
b_{2t}	-	1.00	1.00	1.00	1.00

Table 2.14: Optimal solution for the BHP using Example 2.2

2.2.6 Conversion Problem

With reference to Dannoura and Sakurai [1998], Mohr and Schmidt [2013] and Mohr et al. [2014], assume the investor owns a positive quantity of shares of A_1 at time instant $t = 0$. The aim is to convert these shares until $t = T$ into shares of A_2 such that the terminal wealth W_T is maximized. This problem is called *conversion problem* (CP) and is distinguished into two cases. In the one case, *uni-directional CP* (uniCP), it is only allowed to convert into one direction, e.g., A_1 into A_2 but not A_2 into A_1. In other words A_1 can only be sold and A_2 can only be bought. The other case, *bi-directional CP* (biCP), allows to convert in both directions at all time instants with $t = 0, ..., (T-1)$, e.g., A_1 into A_2 and vice versa.

t	0	1	2	3	$T=4$
q_{1t}	50.00	50.00	50.00	50.00	50.00
q_{2t}	10.00	20.00	9.00	30.00	11.00
q_{3t}	10.00	5.00	9.00	4.00	8.00
$\mathbf{1}_t$	1	0	0	0	1
y_{1t}	-2.0000	0.0000	0.0000	0.0000	0.0000
y_{2t}	10.0000	0.0000	0.0000	0.0000	0.0000
y_{3t}	0.0000	0.0000	0.0000	0.0000	0.0000
Y_{1t}	0.0000	0.0000	0.0000	0.0000	0.0000
Y_{2t}	10.0000	10.0000	10.0000	10.0000	10.000
Y_{3t}	0.0000	0.0000	0.0000	0.0000	0.0000
W_t	100.00	200.00	90.00	300.00	110.00
b_{1t}	-	0.00	0.00	0.00	0.00
b_{2t}	-	1.00	1.00	1.00	1.00
b_{3t}	-	0.00	0.00	0.00	0.00

Table 2.15: Optimal solution for the BHP using Example 2.3

In any CP, if the investor does not convert all remaining shares of A_1 until time instant $t = (T-1)$, then the conversion is fully executed in $t = T$. The remaining shares of A_1 at time instant T are determined and, if there are still any available, sold by

$$y_{1T} = -Y_1^{init} - \sum_{t=0}^{T-1} y_{1t}. \tag{2.17}$$

By rearranging Equation (2.17) to Y_1^{init} and adding it into the Mathematical Program 1, containing the GPSP with $m = 2$ assets, then the mathematical program is changing to Mathematical Program 7 and describes the biCP - the objective function shortens. The number of transactions depends only on the value of the conversion rate $\frac{q_{1t}}{q_{2t}}$ for $t = 0, .., T$.

Mathematical Program 7 - biCP

Given:	$q_{i0},...,q_{iT}$	$, i=1,2$	asset prices
	Y_i^{init}	$, i=1,2$	shares to be owned of A_i at $t=0$
Find:	$y_{10},...,y_{1T}$		shares to convert of A_1 into A_2
max	$W_T = q_{2T}\left(Y_2^{init} + \sum_{t=0}^{T} -y_{1t}\frac{q_{1t}}{q_{2t}}\right)$		
s. t.	(I) $\sum_{t=0}^{\tau} y_{1t} \geq -Y_1^{init}$		$, \tau=0,...,T$
	(II) $\sum_{t=0}^{\tau} -y_{1t}\frac{q_{1t}}{q_{2t}} \geq -Y_2^{init}$		$, \tau=0,...,T$
	(III) $y_{1t} \in \mathbb{R}$		$, t=0,...,T$

The terminal wealth for the biCP is always equal to the one of the GPSP with $m=2$. Mathematical Program 7 is one optimal solution of Mathematical Program 1. This is because, $y_{10},...,y_{1(T-1)}$ is always equal with the GPSP with $m=2$ and y_{1T} does not effect the terminal wealth W_T in the absence of transaction costs.

If the CP is limited to uni-directional conversion then A_1 can only be sold. Thus, the value range of y_{1t} is limited to

$$y_{1t} \leq 0 \tag{2.18}$$

what is equal to $y_{2t} \geq 0$. Taking this limitation into account, Mathematical Program 7 is extended. It follows Mathematical Program 8 which describes the uniCP.

Mathematical Program 8 - uniCP

Given:	$q_{i0},...,q_{iT}$	$, i=1,2$	asset prices
	Y_i^{init}	$, i=1,2$	shares to be owned of A_i at $t=0$
Find:	$y_{10},...,y_{1T}$		shares to convert of A_1 into A_2
max	$W_T = q_{2T}\left(Y_2^{init} + \sum_{t=0}^{T} -y_{1t}\frac{q_{1t}}{q_{2t}}\right)$		
s. t.	(I) $\sum_{t=0}^{\tau} y_{1t} \geq -Y_1^{init}$		$, \tau=0,...,T$
	(II) $y_{1t} \leq 0$		$, t=0,...,T$

An example for the application of Mathematical Program 7 is given in Table 2.16 when Example 2.2 is used. To maximize the terminal wealth W_T the investor needs to decide about the trade in A_1 and A_2 for all $t<T$. If the investor did not convert until time instant $t=(T-1)$, then the full conversion is done at time instant $t=T$. The maximum terminal wealth

t	0	1	2	3	$T=4$
q_{1t}	50.00	50.00	50.00	50.00	50.00
q_{2t}	10.00	20.00	9.00	30.00	11.00
Q_{1t}^2	5.00	2.50	5.56	1.67	4.55
Q_{2t}^1	0.20	0.40	0.18	0.60	0.22
y_{1t}	-2.0000	4.0000	-4.0000	13.3333	-13.3333
y_{2t}	10.0000	-10.0000	22.2222	-22.2222	60.6061
Y_{1t}	0.0000	4.0000	0.0000	13.3333	0.0000
Y_{2t}	10.0000	0.0000	22.2222	0.0000	60.6061
W_t	100.00	200.00	200.00	666.67	666.67
b_{1t}	-	0.00	1.00	0.00	1.00
b_{2t}	-	1.00	0.00	1.00	0.00

Table 2.16: Optimal solution for the biCP using Example 2.2

$W_T = 666.67$ is the same as in the solution for the Mathematical Program 1 given in Table 2.5, but differs in y_{1T}. For completeness b_{it} is given as well.

Table 2.17 gives the solution of Mathematical Program 8 when Example 2.2 is used. In this example, the optimal solution is to execute the full conversion at time instant $t = 2$. This results into the optimal terminal wealth $W_T = 122.22$ for the uniCP.

t	0	1	2	3	$T=4$
q_{1t}	50.00	50.00	50.00	50.00	50.00
q_{2t}	10.00	20.00	9.00	30.00	11.00
Q_{1t}^2	5.00	2.50	5.56	1.67	4.55
Q_{2t}^1	0.20	0.40	0.18	0.60	0.22
y_{1t}	0.0000	0.0000	-2.0000	0.0000	0.0000
y_{2t}	0.0000	0.0000	11.1111	0.0000	0.0000
Y_{1t}	2.0000	2.0000	0.0000	0.0000	0.0000
Y_{2t}	0.0000	0.0000	11.1111	11.1111	11.1111
W_t	100.00	100.00	100.00	333.33	122.22
b_{1t}	-	1.00	1.00	0.00	0.00
b_{2t}	-	0.00	0.00	1.00	1.00

Table 2.17: Optimal solution for the uniCP using Example 2.2

2.3 Standard Working Models

It is shown that all considered problems are special types of the PSP. In the literature the specific problems are often presented from a different perspective. As used in the preceding sections, the perspective of an investor who buys and sells shares of assets depending on the current asset prices is often rejected (perspective 1). For example, in the literature of the MLC for the GPSP and CRP the perspective is on the proportions of wealth and the time convention is based on trading periods, cf. [Li and Hoi, 2014] (perspective 2). In contrast, in the literature for the uniCP the perspective is on conversion rates with the time convention time instants, see [Mohr et al., 2014] (perspective 3). It is shown that all three perspectives belong together.

2.3.1 Portfolio Selection Problem

In the literature of the MLC the GPSP is formulated as the maximization of the terminal wealth depending on the proportion of wealth in A_i, i.e., b_{it} with $\sum_{i=1}^{m} b_{it} = 1$ for all trading periods with $t = 1, ..., T$. The wealth of an investor at time instant $t + 1$ (before rebalancing) can be calculated by

$$W_{(t+1)} = \sum_{i=1}^{m} q_{i(t+1)} Y_{it} \tag{2.19}$$

when Y_{it} is the number of shares owned by the investor after rebalancing at time instant t. The wealth of the investor changes between time instant t and $t + 1$ by

$$\frac{W_{(t+1)}}{W_t} = \frac{\sum_{i=1}^{m} q_{i(t+1)} Y_{it}}{\sum_{i=1}^{m} q_{it} Y_{it}} \tag{2.20}$$

in which Equation (2.7) for the current wealth is combined with Equation (2.19) for the subsequent wealth. Next, Equation (2.13) in a generalized form for arbitrary A_i is rearranged to Y_{it}. It is $Y_{it} = \frac{b_{i(t+1)} W_t}{q_{it}}$ and gives the number of shares of A_i which can be bought for a given wealth W_t and proportion b_{it}. Using this in Equation (2.20) it gives:

$$\frac{W_{(t+1)}}{W_t} = \frac{\sum_{i=1}^{m} q_{i(t+1)} \frac{b_{i(t+1)} W_t}{q_{it}}}{\sum_{i=1}^{m} q_{it} \frac{b_{i(t+1)} W_t}{q_{it}}}. \tag{2.21}$$

The asset prices q_{it} in the numerator are substituted by return factors $\frac{q_{i(t+1)}}{q_{it}} = x_{i(t+1)}$:

$$\frac{W_{(t+1)}}{W_t} = \frac{\sum_{i=1}^{m} x_{i(t+1)} b_{i(t+1)} W_t}{\sum_{i=1}^{m} b_{i(t+1)} W_t}. \tag{2.22}$$

Because $\sum_{i=1}^{m} b_{i(t+1)} = 1$ the denominator becomes W_t. Therefore, the wealth of an investor grows by the factor

$$\frac{W_{(t+1)}}{W_t} = \sum_{i=1}^{m} x_{i(t+1)} b_{i(t+1)} \tag{2.23}$$

between time instant t and $t+1$. If the wealth at time instant t is given, then the wealth of the investor is

$$W_{(t+1)} = W_t \sum_{i=1}^{m} x_{i(t+1)} b_{i(t+1)} \tag{2.24}$$

at time instant $t+1$. For example, if it starts with an initial wealth W_0, then at time instant $t = 1$ (end of the first trading period) the investor owns $W_1 = W_0 \sum_{i=1}^{m} x_{i1} b_{i1}$.

Summing up, for any investment horizon T the wealth is

$$W_T = W_0 \sum_{i=1}^{m} x_{i1} b_{i1} \times \sum_{i=1}^{m} x_{i2} b_{i2} \times ... \times \sum_{i=1}^{m} x_{iT} b_{iT} \tag{2.25}$$

and can be rewritten as

$$W_T = W_0 \prod_{t=1}^{T} \sum_{i=1}^{m} x_{it} b_{it}. \tag{2.26}$$

Note that the PSP, CRP, SPSP, SCRP and BHP can be expressed in the same way, e.g., [Kozat and Singer, 2011] and [Li and Hoi, 2014].

2.3.2 Conversion Problem

Online algorithms for the uniCP are investigated in the MLC with a standard perspective, e.g., Mohr et al. [2014]. At time instant $t = 0$ an amount of wealth W_0 is invested in A_1 and must be converted into a second asset A_2 until time instant $t = T$. It is asked for an amount s_t to be converted, which is a fraction of the available amount of wealth converted at time instant t, with $0 \leq s_t \leq 1$ and $\sum_{t=0}^{T} s_t = 1$. Thus, the investor owns $Y_1^{init} = 1$ and $Y_2^{init} = 0$ initial shares at time instant $t = 0$. Including this into Mathematical Program 8 and replacing the asset prices in the sum with the conversion rates Q_{1t}^2 with $t = 0, ..., T$, the problem is rewritten by maximizing

$$W_T = q_{2T} \sum_{t=0}^{T} s_t Q_{1t}^2 \tag{2.27}$$

subject to $\sum_{t=0}^{\tau} s_t \geq 0$ for $\tau = 0, ..., T$ and $s_t \geq 0$ for $t = 0, ..., T$.

2.4 Conclusions

Before concluding and building a classification structure for the considered types of the PSP, regard the following selected remarks. In the literature there exist several contributions related to the PSP taking online and offline algorithms into account. There are a few, which occasionally describe and categorize several types of the PSP with a differentiated focus on the offline perspective. However, none of them brings the different types of the PSP together in a comprehensive and understanding form such that the specific differences are clearly visible. For example, Borodin et al. [2000, pp. 176-180] remark that online algorithms, which solve a trading problem, can be compared with the three types of offline algorithms, i.e., the optimal solution of the BHP, the CRP and the GPSP. The optimal solution of the GPSP is denoted as "true optimal algorithm". In additon, in the survey of Li and Hoi [2014] the optimal solutions of the BHP and CRP belong to the benchmark group of algorithms as well. But, when to use which benchmark remains an open question. In Györfi et al. [2012, p. 82] trading problems are classified depending on the number of decisions that must be made to solve the problem. For static problems an allocation must be determined only once, i.e., one request. This is equal to the BHP. Dynamic problems include more than one decision and promise higher performance, e.g., the GPSP with more than one request. A further breakdown of problems is lacking. El-Yaniv [1998, pp. 33-34 and pp. 44-46] defines the CP as a special case of the PSP. But a connection based on a holistic model approach is not given. In El-Yaniv et al. [2001], the CP is separated into the optimal search and the one-way trading problem. An explanation how they are connected to the PSP is missing. Cover [1991] is comparing different types of offline benchmarks and proves the relation of the performance between the BHP and the CRP.

Based on the literature, the Mathematical Programs 1 – 8 are developed in order to build a classification structure among selected types of the PSP. The application of the mathematical programs is illustrated with Example 2.2 ($m = 2$) and Example 2.3 ($m = 3$). The summarized results of the examples can be found in Table 2.18.

Type of PSP	Mathematical Program	W_T^* $m = 2$	W_T^* $m = 3$
General portfolio selection problem (GPSP)	1 and 2	666.67	2400.00
Constant rebalancing problem (CRP)	3	161.10	314.81
Semi-portfolio selection problem (SPSP)	4	244.44	440.00
Semi-constant rebalancing problem (SCRP)	5	121.49	150.92
Buy-and-hold problem (BHP)	6	110.00	110.00
Bi-directional conversion problem (biCP)	7	666.67	-
Uni-directional conversion problem (uniCP)	8	122.22	-

Table 2.18: Optimal terminal wealth concerning the considered types of the PSP for Example 2.2 ($m = 2$) and 2.3 ($m = 3$)

The optimal terminal wealth of the considered types of the PSP depends on the four conditions presented in Section 2.1 and the number of assets m in the considered market. Table 2.19 shows the employed conditions to generate the specific problem. Missing combinations of the conditions define further types of the PSP. Note, the SPSP is not always equivalent to the BHP because they can differ in the number of time instants where trading is allowed.

Constraint	GPSP	CRP	SPSP	SCRP	BHP	biCP	uniCP
No short-selling condition	x	x	x	x	x	x	x
No consumption condition	x	x	x	x	x	x	x
Trading condition			x	x	x		
Allocation condition		x		x			
Conversion condition						x	x
Direction condition							x

Table 2.19: Constraints of the considered types of the PSP

The presentation of the selected types of the PSP as mathematical programs gives important insights. It allows partially conclusions about the corresponding online problems, especially on the variation of the terminal wealth W_T for arbitrary instances. Particularly noteworthy remarks for the offline solution of a PSP are:

(i) **Number of assets**: The increase of the number of assets in the market results into an increase of the solution space and therefore the optimal terminal wealth increases or remains the same.

(ii) **Trading Condition**: Reducing the number of time instants where trading is allowed during the investment horizon reduces the optimal terminal wealth or it remains the same.

(iii) **Allocation Condition**: When an allocation condition is added the optimal terminal wealth reduces or remains the same.

(iv) **Conversion Condition**: The conversion condition itself does not effect the terminal wealth when transaction costs are ignored.

(v) **Direction Condition**: When adding a direction condition the optimal terminal wealth reduces or remains the same.

As a result, the selected types of the PSP can be summarized as a tree graph which is shown in Figure 2.1. It gives a simplified classification structure which contains a ranking of the optimal solutions for the considered types of the PSP. The root node of the graph is the GPSP and it let expect the greatest terminal wealth but also the largest feasible region of solutions for arbitrary instances of the market. When adding one additional condition to a specific type of a PSP, then this leads to a new PSP represented as a new node in a new level. One can say, the new PSP is a special case of its root problem. This new PSP has always a lower or equal optimal terminal wealth than the root problem.

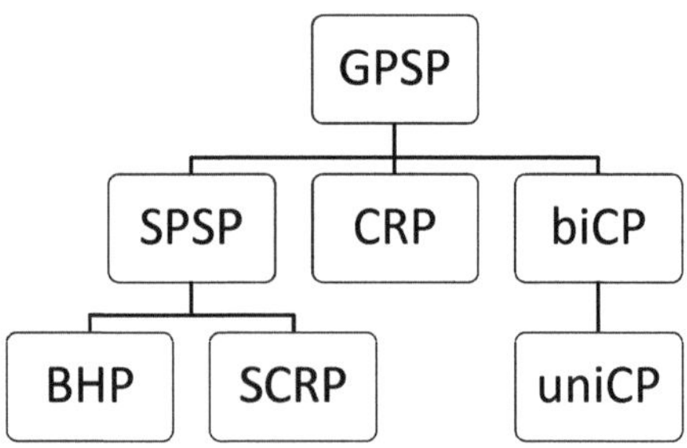

Figure 2.1: Simplified classification structure of selected types of the PSP

One may ask, why build a classification structure for offline problems if the corresponding online algorithms are able to generate contrary results. In practice, the future is assumed to be unknown. Nevertheless, the objective is to construct online algorithms which are *competitive*, to give a performance guarantee related to an offline algorithm for a specific scenario (compare with Albers [2006, p. 1]). For this, the offline problem must be defined exactly and its relation to other problems must be clearly understood.

In fact, an online problem can be only the GPSP, SPSP, biCP and uniCP when the offline perspective is taken away. This is because they can require sequentially actions by the investor. An online algorithm can be compared with an offline algorithm executed by an omniscient adversary. In accordance with Koutsoupias and Papadimitriou [2000], to make the comparison more realistic the power of the adversary can be restricted. This can be done when the adversary can execute only a CRP, SCRP or BHP on the same market instance where the online ALG is applied. The power of the adversary is limited because the CRP, SCRP and BHP require technically only one action at the beginning of the investment horizon. Before developing a new online algorithm it must be clarified, which problem the new algorithm targets and what is the power of the adversary. The provided classification structure gives an orientation for it.

3 Performance Evaluation

Performance evaluation is to make a statement about the behavior of an algorithm based on real data, artificial data or scenarios. In the literature and in practice exist different concepts to evaluate the performance of an algorithm. This chapter presents the main concepts in portfolio selection and the corresponding performance measures motivated by the literature from the finance community and the machine learning community.

3.1 Preliminaries

As derived in Section 2.3.1, a market consists of $A_1, ..., A_m$ assets and $t = 1, ..., T$ trading periods, the latter specifies also the investment horizon of an investor. The price change of asset A_i during trading period t is represented by x_{it}, with $x_{it} \geq 0$. The vector

$$\mathbf{x}_t = (x_{1t}, \ldots, x_{mt}) \tag{3.1}$$

describes the price changes of the assets $A_1, ..., A_m$ during trading period t and the sequence

$$\mathbf{x} = \mathbf{x}_1, \ldots, \mathbf{x}_T \tag{3.2}$$

denotes the entire market. Given an initial wealth W_0, the task is to allocate the wealth among the m assets at the beginning of each trading period t, denoted as PSP. Consider an arbitrary ALG that solves that PSP. This ALG generates at the beginning of each trading period t an allocation vector

$$\mathbf{b}_t = (b_{1t}, \ldots, b_{mt}) \tag{3.3}$$

where b_{it} is a target proportion and quantifies the percentage of wealth, which must be invested in A_i. Given as input $\mathbf{x}$, the sequence

$$\mathbf{b} = \mathbf{b}_1, \ldots, \mathbf{b}_T \tag{3.4}$$

denotes the entire output of the ALG.

Consider the case that the initial wealth W_0 and the market $\mathbf{x}$ is given. An investment into the market $\mathbf{x}$ with an ALG generates a sequence of period wealth $\mathbf{W} = W_0, W_1, ..., W_T$ calculated by

$$W_t = W_0 \prod_{\tau=1}^{t} \sum_{i=1}^{m} x_{i\tau} b_{i\tau} \tag{3.5}$$

for $t = 1, ..., T$.

In the following, the problem of this chapter is stated and some remarks on the performance evaluation are given.

3.1.1 Problem Statement

This chapter deals with two issues. The first issue is, how performance is measured for an ALG, when the entire sequence of period wealth $\mathbf{W}$ based on one market instance $\mathbf{x}$ is given. For this purpose, various performance measures are presented. They are categorized by (i) return on investment, (ii) risk and (iii) risk-adjusted performance. The calculated performance measures can be easily compared with those of another ALG (benchmark) on the same $\mathbf{x}$. Therefore, selected benchmarks motivated by the literature are provided.

The second issue is, how to evaluate the performance of a given ALG with respect to the uncertainty of $\mathbf{x}$. There exists the main concept statistical analysis on the one side, which is based on statistical models and multiple observations of $\mathbf{W}$ with various $\mathbf{x}$. On the other side is the competitive analysis, which is in general free of statistical models and based on specific artificial scenarios.

3.1.2 Efficient Markets Hypothesis

The interpretation of the performance of an ALG for portfolio selection should be carefully done by taking the *efficient markets hypothesis* (EMH) into account, which was formulated by Fama [1970a]. Due to the EMH, capital markets are efficient when the asset prices reflect all historical and current information. A new information is taken into account by an immediate change in the asset price(s). If in the long run an "extra" return on investment is possible due to the exploitation of historical information, then capital markets are inefficient. The EMH is heavily discussed in the financial community [see Malkiel, 2003]. Based on the type of information, the EMH is distinguished into three forms:

(i) **Weak**: The weak EMH claims that the current asset prices reflect all information of historical price movements. Thus, in the long run it is not possible to earn a larger return on investment with the exploitation of historical asset prices without taking more risk at the same time into account.

(ii) **Semi-Strong**: For the semi-strong EMH public information is also considered by the current prices, e.g., fundamental data of an economy or company. For an investor it is in the long run impossible to take advantage of historical price movements and public information.

(iii) **Strong**: The strong EMH takes historical price movements, public information as well as private information into account, i.e., insider information. If the strong EMH holds, then with trading it is not possible to earn a larger return on investment without accepting at the same time more risk.

Summing up, with respect to the EMH, the return on investment of an ALG should always be evaluated in relation to its corresponding risk.

3.1.3 Time Complexity

Besides financial performance, an ALG can be evaluated by the number of operations, which are needed to calculate the output for an input with certain size. Based on the number of operations, the required computational time on a given computer system can be deduced. The number of operations is determined in an average-case or worst-case scenario of the input and is expressed as an asymptotic function by the membership of a complexity class. The commonly used terminology for complexity classes with increasing order of complexity is (i) constant, (ii) logarithmic, (iii) linear, (iv) linearithmic, (v) polynomial, (vi) exponential, (vii) factorial. An ALG with polynomial worst-case time complexity ((i)-(v)) can be calculated in reasonable time for a given input. In contrast, there is no guarantee that an ALG with non-polynomial worst-case time complexity ((vi) and (vii)) can be calculated in reasonable time even with relative small input. Because time is a limited resource, the aim is to develop an ALG with the minimum possible time complexity [see Rosen, 2011, pp. 218-236].

For the PSP, the input is the market $\mathbf{x}$ with size m and T such that the time complexity of an ALG is expressed as a function $O(\cdot) = f(T, m)$. Remarks on the time complexity of algorithms in online portfolio selection can be found for example in [Helmbold et al., 1998] and [Li et al., 2013], where algorithms with polynomial and non-polynomial worst-case time complexity are discussed. Due to limited resources of computational time, the application of algorithms with non-polynomial time complexity can be a problem in practice. Thus, the performance evaluation of an ALG should always take the analysis of time complexity into account. (An overview of time complexities of prominent online algorithms can be found in Table 4.3 on Page 107.)

3.2 Selected Performance Measures

It follows a selection of performance measures. They are the basic measures for all subsequent considerations with respect to the performance evaluation of an ALG that solves the PSP. The classification of performance measures is done by (i) return on investment, (ii) risk and (iii) risk-adjusted performance. Each performance measure can be calculated in hindsight when the entire sequence of period wealth $\mathbf{W}$ is given.

3.2.1 Measures of Return on Investment

A measure of return on investment quantifies the amount of profit of an ALG. They come in a variety of forms. They can be absolute values that show the profit in monetary units, or they can be relative values, which characterizes the amount of profit growth.

Terminal Wealth

The terminal wealth itself is a performance measure and is calculated by

$$W_T = W_0 \prod_{t=1}^{T} \sum_{i=1}^{m} x_{it} b_{it} \tag{3.6}$$

for any output of an ALG when W_0 and $\mathbf{x}$ are fully known. Obviously, $W_T \geq 0$ when $x_{it} \geq 0$ is true and from a simplified perspective a rational investor aims to maximize W_T, e.g., [Cover, 1991], [Borodin et al., 2004] and [Li and Hoi, 2014]. If $W_0 = 1$, then the terminal wealth is sometimes called total return. Alternatively, for arbitrary W_0 the total return is calculated by $\frac{W_T}{W_0}$. According to Modigliani and Modigliani [1997, p. 45], the total return or respectively the terminal wealth is an incomplete measure of performance because it does not incorporate risk.

Exponential Growth Rate

Let W_t be the period wealth at the end of trading period t generated by an ALG. Then,

$$x_t = \frac{W_t}{W_{(t-1)}} \tag{3.7}$$

is the holding period return for trading period $t = 1, ..., T$ showing the increase/decrease of the wealth holding a portfolio $\mathbf{b}_t$ for one single trading period. Then, the performance measure *exponential growth rate*, denoted as μ, can be calculated in various forms - in an additive form [compare to Györfi et al., 2007, p. 507] by

$$\mu = \frac{1}{T}\sum_{t=1}^{T} \ln x_t \tag{3.8}$$

or a multiplicative form [see Cover, 1991, p. 3] by

$$\mu = \ln \sqrt[T]{\prod_{t=1}^{T} x_t}. \tag{3.9}$$

But, the insertion of Equation (3.7) into Equation (3.8) and (3.9) leads to

$$\mu = \frac{1}{T} \ln \frac{W_T}{W_0} \tag{3.10}$$

or the logarithmic geometric mean return

$$\mu = \ln \sqrt[T]{\frac{W_T}{W_0}}. \tag{3.11}$$

The μ is named as *exponential* growth rate because it is easy to show that

$$W_T = W_0 \exp(\mu T) \tag{3.12}$$

and it quantifies how much the wealth increases on average during one trading period, cf. [Cover, 1991] and [Cover and Ordentlich, 1996].

Annual Percentage Yield

With reference to Li et al. [2012, p. 239], the *annual percentage yield* (APY) is calculated by

$$APY = \left(\frac{W_T}{W_0}\right)^{\frac{1}{(\text{Number of years})}} - 1 \tag{3.13}$$

which depends on the number of years corresponding to T trading periods. It takes the compounding effect into account and measures the average wealth increment that an ALG could achieve compounded in one year. Thus, this measure can be also calculated by

$$APY = \exp\left(\mu \times (\text{Trading periods in one year})\right) \tag{3.14}$$

with the number of trading periods from $t = 1, ..., T$ which are required to complete one year.

3.2.2 Measures of Risk

The potential and the extent of negative performance of an ALG is quantified by measures of risk.

Standard Deviation and Variance

According to McMillan et al. [2011, p. 189], the spread out of the exponential growth rate is measured by the *standard deviation* or *variance.* The standard deviation[1] of the exponential growth rate is calculated by

$$\sigma = \sqrt{\frac{1}{T}\sum_{t=1}^{T}\left(\ln x_t - \mu\right)^2} \tag{3.15}$$

and the corresponding variance is

$$\sigma^2 = \frac{1}{T}\sum_{t=1}^{T}\left(\ln x_t - \mu\right)^2 . \tag{3.16}$$

From this perspective the standard deviation is a measure of risk and is expressed in the same unit as the exponential growth rate. It follows that

[1] Note that if the standard deviation is merely used as a performance measure, it is not necessarily to be distinguished in sample and population standard deviation, because for large T the difference becomes extremely small.

this measure (and the variance) is always non-negative and a low value indicates holding period returns x_t at time $t = 1, ..., T$ very close to the exponential growth rate, i.e., low risk. In contrast, a high value means that the holding period returns spread out far from that exponential growth rate, i.e., high risk. The corresponding *annualized standard deviation* (ASTDV) is quantified by

$$ASTDV = \sigma\sqrt{(\text{Trading periods in one year})} \tag{3.17}$$

and depends on the required number of trading periods from $t = 1, ..., T$ to complete one year. The ASTDV as performance measure is in use for several investigations concerning online portfolio selection, e.g., [Borodin et al., 2004], [Agarwal et al., 2006] and [Li et al., 2012]. The drawback of the standard deviation as a risk measure is that upside and downside movements of the wealth are taken into account. But, upside movements are usually desired by an investor and therefore sometimes not considered as risk [compare with Estrada, 2006].

Maximum Drawdown

According to Li et al. [2012, p. 239], the *maximum drawdown* (MDD) is a measure for the downside risk, i.e., a risk measure which takes only loss movements of wealth into account. Following Magdon-Ismail et al. [2004, p. 147], it is the largest drop from the peak to the subsequent bottom of the period wealth. The drawdown for trading period t is calculated by $DD_t = \max_{\tau=0,...,t} W_\tau - W_t$ such that the maximum drawdown is

$$MDD = \max_{t=0,\dots,T}\left(\max_{\tau=0,\dots,t} W_\tau - W_t\right). \tag{3.18}$$

From reasons of comparability it is recommended to normalize the period wealth such that the total return is employed ($W_0 = 1$).

For an investor it seems to be beneficial to measure the drawdown in percentage drop from the peak. According to Vecer [2006, pp. 91,92], this is the relative drawdown and is calculated for each trading period t by $RDD_t = 1 - \frac{W_t}{\max_{\tau=0,\dots,t} W_\tau}$. Thus, the *maximum relative drawdown* (MRDD) is quantified by

$$MRDD = \max_{t=0,\dots,T}\left(1 - \frac{W_t}{\max_{\tau=0,\dots,t} W_\tau}\right). \tag{3.19}$$

In contrast to the standard deviation, the MDD and MRDD consider only effects for a decrease of wealth, i.e., negative risk. Positive movements of wealth are not considered in the risk calculation.

3.2.3 Measures of Risk-adjusted Performance

According to [Modigliani and Modigliani, 1997, p. 45], an investor is able to obtain a higher return on investment, if a greater level of risk is accepted. To indicate if an ALG achieves a higher terminal wealth W_T due to luck and a greater risk level, the performance evaluation should take measures for the risk-adjusted performance into account.

Reward-to-Variability Ratio

The most commonly used risk-adjusted performance measure is the *reward-to-variability ratio* (RVR), motivated by Sharpe [1966] and often called Sharpe ratio. In principle, the percentage yield R_f for a risk-free asset is subtracted from the percentage yield for the ALG in the same considered investment horizon (year). The remaining excess yield is put into ratio with the standard deviation of that yield. Thus, the annual RVR is calculated by

$$RVR = \frac{APY - R_f}{ASTDV} \tag{3.20}$$

with reference to Li et al. [2012, p. 239]. For simplification, R_f is sometimes assumed to be zero [e.g., Campbell et al., 2001, p. 1794]. Basically, a higher RVR is better because it indicates a higher return on investment per unit of risk. The RVR can be also negative. Then, it indicates a worse performance than the risk-free asset, but the interpretation of that negative value needs to be done carefully [for more details see Mayo, 2010, p. 222].

Drawdawn Ratio

A similar risk-adjusted performance measure as the RVR is the *drawdown ratio* (DDR), which is often called Calmar ratio. Instead of the standard deviation the MDD is employed. The drawback of taking positive upside movements of wealth into account is eliminated. There is no assumption concerning the risk-free asset required. The formula for the DDR is

$$DDR = \frac{APY}{MDD} \tag{3.21}$$

where a higher DDR indicates a better performance [Li et al., 2012, p. 239].

3.3 Selected Benchmarks

If the performance of an ALG is compared with another ALG, then this second ALG is called benchmark. A benchmark can be any online or offline ALG, but mainly it is a simplified ALG, e.g., (i) buy a portfolio only once and hold it for the entire investment horizon (buy-and-hold ALG) or (ii) buy at each trading period a portfolio and ignore historical data (non-learning ALG). To support the following considerations, let

$$\mathfrak{B}_m = \left(\mathfrak{b} \in \mathbb{R}^m : \mathfrak{b}_i \geq 0, \sum_{i=1}^{m} \mathfrak{b}_i = 1 \right) \tag{3.22}$$

be a simplex, containing all possible allocations of wealth on m assets [cf. Cover, 1991, p. 2].

In general, there exists an infinite number of allocations within the simplex $\mathfrak{B}_m$, because $\mathfrak{b}_i \in \mathbb{R}$. But, for reasons of simplification the continuous simplex can be discretized and the number of allocations within the simplex is reduced to a finite number. Let a be a multiple factor with $0 < a \leq 1$ and $\frac{1}{a} \in \mathbb{N}^{+}$. In addition, assume each $\mathfrak{b}_i$ is a multiple of a, i.e., $\frac{\mathfrak{b}_i}{a} \in \mathbb{N}^0$. An investor has to allocate $\frac{1}{a}$ times a percentage of wealth a on m assets. At the end, $\frac{1}{a} \times a = 100\%$ of the wealth is allocated on the m assets. The investor is able to perform repetitive selection of one asset. It is questionable how many possible allocations are in a simplex $\mathfrak{B}_m$ when a and m are given. It is obvious, this is a combinatorial problem with repetition where the order of selection does not matter [cf. Rosen, 2011, p. 427]. Thus,

$$\Omega = \frac{(m + \frac{1}{a} - 1)!}{\frac{1}{a}!\,(m-1)!} = \binom{m + \frac{1}{a} - 1}{\frac{1}{a}} \tag{3.23}$$

quantifies the number of different allocations when the simplex $\mathfrak{B}_m$ is discretized. Table 3.1 presents the number of allocations for various a and m. For instance Cover [1991, p. 20] is using $a = 0.05$ with $m = 2$ and illustrates also 21 possible allocations. The number of allocations essentially increases with the reduction of a and enlarging of m. More precisely, the number of allocations Ω increases exponentially as shown in Lemma A.3 in the appendix. It follows a selection of benchmarks using a continuous simplex $\mathfrak{B}_m$. They can be easily modified using a discretized simplex. The

a	$m=2$	$m=3$	$m=4$	$m=5$
1	2	3	4	5
0.5	3	6	10	15
0.1	11	66	286	1001
0.05	21	231	1771	10626

Table 3.1: Number of possible allocations in the market for four various a and four different m

provided benchmarks are mainly motivated by the literature of the FC but combined with the perspective of the MLC.

3.3.1 Offline Benchmarks: Buy-and-Hold

The most simple benchmarks are from the type of *buy-and-hold portfolio* (BH). For this, assume the terminal wealth for the benchmark is calculated by

$$W_T^{BH}(\mathfrak{b}) = W_0 \sum_{i=1}^{m} \mathfrak{b}_i \prod_{t=1}^{T} x_{it} \tag{3.24}$$

such that the wealth is only allocated at the beginning of trading period $t=1$. The period wealth for arbitrary t would be calculated by $W_t^{BH}(\mathfrak{b}) = W_0 \sum_{i=1}^{m} \mathfrak{b}_i \prod_{\tau=1}^{t} x_{i\tau}$.

Best Asset Portfolio

The best BH in hindsight concerning the terminal wealth is defined as

$$\mathfrak{b}^{BA} = \arg\max_{\mathfrak{b}\in\mathfrak{B}_m} W_T^{BH}(\mathfrak{b}) \tag{3.25}$$

and consists always of one single asset, i.e., the *best asset portfolio* (BA). This is also the asset with the highest exponential growth rate [refering to Li and Hoi, 2014, p. 7].

Best Variance Buy-and-Hold Portfolio

There exists at least one allocation of wealth $\mathfrak{b}$ which generates the lowest risk in hindsight. Let **Var** $[\cdot]$ be a risk function according to Equation (3.16), then this portfolio is defined as

$$\mathfrak{b}^{BVBH} = \arg\min_{\mathfrak{b}\in\mathfrak{B}_m} \mathbf{Var}\left[W_0, W_1^{BH}(\mathfrak{b}), ..., W_T^{BH}(\mathfrak{b})\right] \tag{3.26}$$

and denoted as the *best variance buy-and-hold portfolio* (BVBH). The BVBH is in accordance with the theory of [see Markowitz, 1952]. In principle, the "minimum variance portfolio" roughly corresponds to the BVBH when it would be determined ex-post on a concrete market, e.g., [Clarke et al., 2006].

Best Reward-To-Variability Buy-and-Hold Portfolio

In addition, there exists one portfolio with the best relation between risk and the return on investment. Let $\mathbf{RVR}[\cdot]$ be a function to measure the performance according to Equation (3.20). Then,

$$\mathfrak{b}^{BRVBH} = \arg\max_{\mathfrak{b}\in\mathfrak{B}_m} \mathbf{RVR}\left[W_0, W_1^{BH}(\mathfrak{b}), ..., W_T^{BH}(\mathfrak{b})\right] \tag{3.27}$$

is the *best reward-to-variability buy-and-hold portfolio* (BRVBH). The BRVBH roughly corresponds to an ex-post determined "market portfolio" of the FC for a concrete market, e.g., [Sharpe, 1966] and [Clarke et al., 2006].

3.3.2 Offline Benchmarks: Constant Rebalancing

A portfolio, which is rebalanced at the beginning of each trading period for $t = 1, ..., T$ and fulfills the constraint $\mathbf{b}_{(t+1)} = \mathbf{b}_t$ for $t = 1, ..., (T-1)$ is called *constant rebalancing portfolio* (CR). Thus,

$$W_T^{CR}(\mathfrak{b}) = W_0 \prod_{t=1}^{T} \sum_{i=1}^{m} x_{it}\mathfrak{b}_i \tag{3.28}$$

is the terminal wealth for a given proportion $\mathfrak{b}$, which is hold constant for $t = 1, ..., T$. The corresponding period wealth for $t = 1, ..., T$ can be determined by $W_t^{CR}(\mathfrak{b}) = W_0 \prod_{\tau=1}^{t} \sum_{i=1}^{m} x_{i\tau}\mathfrak{b}_i$. It can be shown that constant rebalancing algorithms are theoretical and practical beneficial in comparison to buy-and-hold algorithms, e.g., see Cover and Gluss [1986] and Cover [1991]. The best portfolio allocation is chosen in hindsight depending on one specific performance measure. In accordance with the offline benchmarks for the BH the following benchmarks with constant rebalancing are defined.

Best Constant Rebalancing Portfolio

Let

$$\mathfrak{b}^{BCR} = \arg \max_{\mathfrak{b} \in \mathfrak{B}_m} W_T^{CR}(\mathfrak{b}) \tag{3.29}$$

be the *best constant rebalancing portfolio* (BCR) concerning the terminal wealth as the performance measure. This portfolio is equal to the portfolio with the highest exponential growth rate in hindsight. It can be shown that this exponential growth rate will be greater or equal than all possible allocations with a BH [Cover, 1991, pp. 1-6]. Remark, that any ALG which provably tracks this benchmark is called "universal". For example, see the introduced algorithms of Cover [1991] and Helmbold et al. [1998] (see also Section 3.5.5).

Best Variance Constant Rebalancing Portfolio

In addition, let

$$\mathfrak{b}^{BVCR} = \arg \min_{\mathfrak{b} \in \mathfrak{B}_m} \mathbf{Var}\left[W_0, W_1^{CR}(\mathfrak{b}), ..., W_T^{CR}(\mathfrak{b})\right] \tag{3.30}$$

be the CR with the lowest risk, in terms of variance of the exponential growth rate. The motivation of such a benchmark is based on [Agarwal et al., 2006, p. 15], where a mean-variance CR for a given ALG and market $\mathbf{x}$ is searched. This CR achieves the same exponential growth rate as this ALG, but has a minimum variance.

Best Reward-To-Variability Constant Rebalancing Portfolio

For the sake of completeness, it makes sense to ask for a benchmark with the best relationship between risk and return on investment. Thus, let

$$\mathfrak{b}^{BRVCR} = \arg \max_{\mathfrak{b} \in \mathfrak{B}_m} \mathbf{RVR}\left[W_0, W_1^{CR}(\mathfrak{b}), ..., W_T^{CR}(\mathfrak{b})\right] \tag{3.31}$$

be the *best reward-to-variability constant rebalancing portfolio* (BRVCR).

3.3.3 Online Benchmarks

An alternative to offline benchmarks are online benchmarks. They neglect the drawback of calculations in hindsight and are therefore more realistic. In the basic form they do not take any past information into account. In this case, the market behavior $\mathbf{x}_{t'}$ with $t' < t$ is not taken into account and has no effect on the calculation of $\mathbf{b}_t$, i.e., non-learning benchmark. In contrast, a learning benchmark tries to exploit past information in order to achieve a target, e.g., minimizing risk or maximizing return on investment. In principle, any online algorithm can be a benchmark algorithm. Two non-learning benchmarks are provided, the naive diversification and the random portfolio.

Naive Diversification

An often used benchmark in theory and practice is the *naive diversification* (ND). The wealth is diversified into m equal parts with $\mathfrak{b}_i = \frac{1}{m}$, i.e.,

$$\mathfrak{b}^{ND} = \left(\frac{1}{m}, \cdots, \frac{1}{m}\right). \tag{3.32}$$

According to DeMiguel et al. [2009], an ALG based on $\mathfrak{b}^{ND}$ seems to be beneficial in comparison to a large variety of optimization approaches. Following Li et al. [2011, p. 5], combining $\mathfrak{b}^{ND}$ with Equation (3.24) results into a BH and specifies the average development of the market, denoted as market strategy or *uniform BH* (UBH). Beating the benchmark UBH means "beating the market". In contrast, when $\mathfrak{b}^{ND}$ is combined with constant rebalancing (Equation (3.28)) then the generated portfolio is denoted as *uniform CR* (UCR).

Random Portfolios

Assume $\mathfrak{b}^{RAND}$ is a *random portfolio* (RAND). By combining RAND with various types of the PSP a variety of random solutions can be generated. These can be used as non-learning benchmarks. (i) The combination with Equation (3.24) generates a random solution for the BHP. (ii) Including the random allocation into Equation (3.28) brings out a random solution for the CRP. (iii) If it is combined with Equation (3.6) and $\mathfrak{b}^{RAND}$ is resampled for each $t = 1, ..., T$, then a complete random solution for the GPSP is produced. The idea of using a RAND as a benchmark is that a learning ALG is mainly based on a model that takes past information into account. If the ALG is suitable for solving a specific type of the PSP with respect to a specific objective function, then the ALG should be at least as good as the majority of RANDs, which ignore past information. More details of using a RAND as a benchmark can be found in Burns [2006] and Lisi [2011].

Ishijima [2001, pp. 6-8] provides an algorithm to generate a RAND with uniform sampling for m assets in a continuous simplex $\mathfrak{B}_m$: Let

$u_1, ..., u_{(m-1)}$ be a sorted (increasing) sequence of $(m-1)$ random numbers from the standard uniform distribution to generate one random portfolio in the simplex $\mathfrak{B}_m$. Then, the random portfolio $\mathfrak{b}^{RAND}$ is determined by

$$\begin{aligned} \mathfrak{b}_1^{RAND} &= u_1 \\ \mathfrak{b}_i^{RAND} &= u_i - u_{(i-1)} \quad , i = 2, ..., (m-1) \\ \mathfrak{b}_m^{RAND} &= 1 - u_{(m-1)}. \end{aligned} \tag{3.33}$$

3.4 Statistical Analysis

Statistical analysis is a concept which can be used to evaluate the performance of an ALG based on real or artificial data. In principle, the ALG is executed and observed multiple times on that data and a set of observations for a specific performance measure arose. Based on the set of observations an indication on the behavior of the ALG can be stated by the use of statistical measures and hypothesis testing.

More formal, let $\mathbf{x}^{(1)}, ..., \mathbf{x}^{(N)}$ be N different markets with unique m and T. In addition, let $Perf(\cdot)$ be a function for an arbitrary performance measure, as for example described in Section 3.2. Then, $X^{(j)} = Perf(ALG, \mathbf{x}^{(j)})$ represents the performance of the ALG on the specific market instance $\mathbf{x}^{(j)}$. The performance of the ALG is measured on N various markets such that $X^{(1)}, ..., X^{(N)}$ observations of the performance measure for the ALG exist. If N is sufficiently large, then statistical analysis can be applied to make a general statement on the performance of ALG with respect to the considered observations.

3.4.1 Selected Statistical Measures

For a better understanding, assume X is a random variable, which can be observed by $X = \{X^{(1)}, ..., X^{(N)}\}$ where an empirical distribution $\hat{\mathbf{F}}$ can be drawn, i.e., $X \sim \hat{\mathbf{F}}$. Based on such distribution statistical measures can be calculated. Selected measures are taken from Rachev et al. [2010, chap. 3]:

Arithmetic Mean

The arithmetic mean is the quantity of the sum of all values $X^{(1)}, ..., X^{(N)}$ divided by the number of values. This number can be either the population size or sample size, denoted as N. Assuming that the population can be completely observed then the arithmetic mean is defined as

$$\mathbf{E}[X] = \hat{\mathbf{E}}[X] = \frac{1}{N}\sum_{j=1}^{N} X^{(j)}. \tag{3.34}$$

But note, technically N would be different for sample mean $\hat{\mathbf{E}}[X]$ and population mean $\mathbf{E}[X]$. This statistical measure reveals the center of the observations. Clearly, it is the only single number where the distance to all observations sum up to zero $(\sum_{j=1}^{N}(X^{(j)} - \mathbf{E}[X]) = 0)$. In other words, when selecting observations on the performance measure for an ALG randomly one has to expect, on average, the arithmetic mean.

Standard Deviation

The standard deviation quantifies the variation of observations. It is measured by the squared deviation of each observation from the arithmetic mean. Larger deviations have higher effect on the standard deviation than smaller deviations because of the square. Formally, the population standard deviation is calculated by

$$\mathbf{SD}[X] = \sqrt{\frac{1}{N}\sum_{j=1}^{N}\left(X^{(j)} - \mathbf{E}[X]\right)^2} \tag{3.35}$$

and the sample standard deviation by

$$\hat{\mathbf{SD}}[X] = \sqrt{\frac{1}{N-1}\sum_{j=1}^{N}\left(X^{(j)} - \mathbf{E}[X]\right)^2} \tag{3.36}$$

where the only difference is that the sum of squared deviation is divided by either N or $(N-1)$. Obviously, the divergence between sample and population standard deviation shrinks almost to zero with large N, i.e., $\lim_{N\to\infty}\left(\mathbf{SD}[X] - \hat{\mathbf{SD}}[X]\right) = 0$. The scale of the standard deviation is the same as the original scale of the observations. If selecting at random an observation where the standard deviation is high (low), then, on average, one can expect to be far (close) from the arithmetic mean.

Skewness

The skewness indicates whether the observations are distributed symmetrically around the arithmetic mean or not. For help, consider the distribution

$\hat{\mathbf{F}}$ and separate it into two halves, exact in the middle. Ideally, if the skewness is zero, then observations are symmetrical distributed, i.e., $\hat{\mathbf{F}}$ is symmetric. If the skewness is lower (larger) zero, then the observations are negative (positive) skewed. In other words, most of the observations are on the right (left) half of the distribution $\hat{\mathbf{F}}$. The population skewness is calculated by

$$\mathbf{SKEW}[X] = \frac{1}{N}\sum_{j=1}^{N}\left(\frac{X^{(j)} - \mathbf{E}[X]}{\mathbf{SD}[X]}\right)^3 \tag{3.37}$$

and the sample skewness $\mathbf{SK\hat{E}W}[X]$ is calculated identically, but uses $\hat{\mathbf{E}}[X]$ and $\mathbf{S\hat{D}}[X]$. Thus, if one is selecting an observation at random where the observations are negatively (positively) skewed, then one can expect, on average, a value on the right (left) half of the distribution.

Kurtosis

The kurtosis is a measure for the peakedness of the distribution $\hat{\mathbf{F}}$ in comparison to the statistical important normal distribution. For a normal distribution the kurtosis is 3. Thus, a value lower (greater) than 3 indicates that less (more) observations are at the tails of the distribution as it would be if $\hat{\mathbf{F}}$ would be a normal distribution. In other words, a kurtosis lower (greater) than 3 is often called thin (fat) tailed distribution. Thin tailed distributions gathering more observations around the arithmetic mean. In fat tailed distributions occur more extreme values than in the normal distribution. The kurtosis is calculated by

$$\mathbf{KURT}[X] = \frac{1}{N}\sum_{j=1}^{N}\left(\frac{X^{(j)} - \mathbf{E}[X]}{\mathbf{SD}[X]}\right)^4 \tag{3.38}$$

and the sample kurtosis $\mathbf{KU\hat{R}T}[X]$ is calculated identically, but using $\hat{\mathbf{E}}[X]$ and $\mathbf{S\hat{D}}[X]$. If one is selecting at random one observation, if all observations imply a kurtosis lower (greater) 3, then, on average, one can expect more (less) a value around the arithmetic mean than when the observations would be normal distributed.

Quantile, Minimum and Maximum

A quantile divides ordered observations into two parts. The lower (left) part represents α percentage and the upper (right) part $(1 - \alpha)$ percentage of the observations. Formally, the α-quantile is expressed by

$$\hat{\mathbf{Q}}_\alpha[X]{=}\mathbf{Q}_\alpha \text{ where } \frac{1}{N}|\{j|X'^{(j)}{\leq}\mathbf{Q}_\alpha\}|{\geq}\alpha \text{ and } \frac{1}{N}|\{j|X'^{(j)}{\geq}\mathbf{Q}_\alpha\}|{\leq}1{-}\alpha \tag{3.39}$$

where $X'^{(1)},...,X'^{(N)}$ is the ordered sequence (increasing) of the observations $X^{(1)},...,X^{(N)}$. One can say, if an observation is taken at random, then the taken observation will be with a probability of α lower or equal and with a probability of $1-\alpha$ greater than the value of the $\hat{\mathbf{Q}}_\alpha[X]$. The quantile with $\alpha = 0.5$ separates the observations into two parts where the lower part has the same size as the upper part and is called median, i.e., $\mathbf{MED}[X] = \hat{\mathbf{Q}}_{0.5}[X]$. In this context, one may ask for the minimum and maximum value of observations. Let $\mathbf{MIN}[X] = X'^{(1)}$ be the minimum value and $\mathbf{MAX}[X] = X'^{(N)}$ be the maximum value.

3.4.2 Hypothesis Testing

When the performance of an ALG is measured on various market instances, one may ask whether the performance of one ALG is significantly "better" than a target value or another ALG. Referring to Kanji [2006, pp. 1-4], when sufficient observations are collected, it is possible to verify such statement concerning the general behavior or property of the considered random variable, i.e., the performance measure for an ALG. For this reason, two corresponding hypothesis are formulated and combined with statistical measures such that a hypothesis system can be executed. The method of applying and executing a hypothesis system to validate a statement is called hypothesis testing and is usually separated into five steps:

(i) **Formulate Hypothesis**: The statement is formulated in terms of two corresponding hypothesis, i.e., the hypothesis system H which contains the null hypothesis H_0 and the alternative hypothesis H_1. H_0 represents the situation of "status quo" and H_1 the counter range of the situation, which is formulated to diagnose the initial statement.

(ii) **Calculate Test Statistic**: A mathematical function, the test statistic $\mathcal{T}$, is calculated and is only based on the numerical observations. The $\mathcal{T}$ behaves differently, i.e., either H_0 is true or H_1 is true. In addition, $\mathcal{T}$ is calculated under the assumption that H_0 is true.

(iii) **Choose Critical Region**: A critical region must be chosen which will strongly indicates that H_0 is rejected and therefore H_1 is true. There are three types of critical regions: (1) both-sided, H_0 is rejected when $\mathcal{T}$ is *either* less or equal or greater *or* equal than a critical value (2) left-sided, H_0 is rejected when $\mathcal{T}$ is less or equal than a critical value and (3) right-sided, H_0 is rejected when $\mathcal{T}$ is greater or equal than a critical value.

(iv) **Choose Significance Level**: The size of the critical region is quantified by the significance level α. Based on α the bound of the critical region is determined. It is usually set to a value between 1 and 10 percent. In other words, α quantifies the risk to reject H_0 when it is in fact true, i.e., the "type I error". If H_0 is not rejected, but in fact it is false and H_1 is true, then this is called "type II error".

(v) **Determine Actual Significance Level**: Optionally, quantify where $\mathcal{T}$ is in the critical region, denoted as actual significance level. It provides useful information on how close H_0 is rejected or not.

In the following a selection of hypothesis systems is provided, which is useful to evaluate the performance of an ALG. The combination of one hypothesis system H with a specific test statistic $\mathcal{T}$ is called a statistical test. In the literature a large number of predefined statistical tests exist. Because, a lot of these require the normal distribution of observations or at least for the underlying population, testing for normal distribution is regarded. Testing is distinguished into testing with one sample and testing with two samples. In testing with one sample the performance of an ALG is evaluated against a given target value. In contrast, testing with two samples considers the evaluation of the performance of ALG against a benchmark. Statistical tests with and without the requirement of normal distribution of observations are provided, i.e., parametric and nonparametric statistical tests.

Testing for Normal Distribution

To test for normal distribution of observations it is assumed that the observations are normal distributed until H_0 is rejected. Thus, the hypothesis system H is formulated as

$$H = \begin{cases} H_0: & X \text{ is normal distributed} \\ H_1: & X \text{ is not normal distributed} \end{cases} \tag{3.40}$$

such that it is a right-sided type of critical region. The system must be solved by a given test statistic which is greater or equal than a critical value. The *Jarque Bera test* (JBT) is one possible test with the test statistic

$$\mathcal{T}^{JBT} = \frac{N}{6}\left((\mathbf{SD}[X])^2 + \frac{(\mathbf{KURT}[X]-3)^2}{4}\right) \tag{3.41}$$

which must be compared with a critical value taken from the chi-squared distribution χ^2 with degree of freedom 2 [see Rachev et al., 2010, pp. 571-572]. Critical values for various significance levels α are provided in Table 3.2 [compare to Kanji, 2006, p. 195]. If $\mathcal{T}^{JBT}$ is greater or equal than the critical value, then H_0 must be rejected and H_1 is true, i.e., X is not normal distributed.

α	0.1	0.05	0.025	0.005
critical value	4.61	5.99	7.38	9.21

Table 3.2: Critical values for various significance levels α of the chi-squared distribution χ^2 with degree of freedom 2

Testing with One Sample

In testing with one sample a statement concerning the statistical measure of one observed random variable X is verified. Exemplary, consider as statistical measure the arithmetic mean $\hat{\mathbf{E}}[X]$, which is compared to a given threshold value $\mathbf{E}_0$. As an example regard the following statement: On average, the exponential growth rate of an ALG $\hat{\mathbf{E}}[\mu]$ is greater than $\mathbf{E}_0 = 0$, i.e., $\hat{\mathbf{E}}[\mu] > 0$.

There are three cases to formulate a statement which results in a hypothesis system. When verifying $\hat{\mathbf{E}}[X] \neq \mathbf{E}_0$, it is hypothesis testing with a both-sided critical region and is formulated by the system

$$H^{both-sided} = \begin{cases} H_0: & \hat{\mathbf{E}}[X] = \mathbf{E}_0 \\ H_1: & \hat{\mathbf{E}}[X] \neq \mathbf{E}_0 \end{cases}, \tag{3.42}$$

but verifying $\hat{\mathbf{E}}[X] < \mathbf{E}_0$ is formulated by

$$H^{left-sided} = \begin{cases} H_0: & \hat{\mathbf{E}}[X] \geq \mathbf{E}_0 \\ H_1: & \hat{\mathbf{E}}[X] < \mathbf{E}_0 \end{cases} \tag{3.43}$$

and verifying $\hat{\mathbf{E}}[X] > \mathbf{E}_0$ is expressed by

$$H^{right-sided} = \begin{cases} H_0: & \hat{\mathbf{E}}[X] \leq \mathbf{E}_0 \\ H_1: & \hat{\mathbf{E}}[X] > \mathbf{E}_0 \end{cases}. \tag{3.44}$$

$H^{both-sided}$, $H^{left-sided}$ and $H^{right-sided}$ are solvable by one test statistic, which must be chosen depending on the assumptions on the observations. If the standard deviation of the underlying population is known, then use the test statistic of the *one sample z-test* (OZT), calculated by

$$\mathcal{T}^{OZT} = \frac{\hat{\mathbf{E}}[X] - \mathbf{E}_0}{\mathbf{SD}[X]/\sqrt{N}}, \tag{3.45}$$

but if the population standard deviation is not known, then use the test statistic of the *one sample t-test* (OTT), calculated by

$$\mathcal{T}^{OTT} = \frac{\hat{\mathbf{E}}[X] - \mathbf{E}_0}{\hat{\mathbf{SD}}[X]/\sqrt{N}}. \tag{3.46}$$

When using these test statistics it must be assumed that the population of X is normal distributed. If not, the results give at least a good approximation. For the OZT, compare the test statistic with the critical value of the standard normal distribution (see Table 3.3). In contrast, in the OTT the test statistic is compared with the critical value of the Student's t-distribution with $N-1$ degrees of freedom. Note, with increasing N the results of $\mathcal{T}^{OTT}$ converges always to $\mathcal{T}^{OZT}$. In addition, the Student's t-distribution converges always to the standard normal distribution with increasing degrees of freedom. Thus, if N is sufficiently large, then $\mathcal{T}^{OZT}$ can be always preferred.

Decisively, let be z_α the critical value of the standard normal distribution (e.g., see Table 3.3). Hence, if $|\mathcal{T}^{OZT}| \geq z_{\alpha/2}$, then reject H_0 for $H^{both-sided}$, if $\mathcal{T}^{OZT} \leq -z_\alpha$, then reject H_0 for $H^{left-sided}$ and if $\mathcal{T}^{OZT} \geq z_\alpha$, then reject H_0 for $H^{right-sided}$. The formulations for the rejection when using $\mathcal{T}^{OTT}$ and applying the Student's t-distribution are identical. The previous explanations correspond to Kanji [2006].

α	0.1	0.05	0.025	0.005
critical value (z_α)	1.28	1.64	1.96	2.58
critical value ($z_{\alpha/2}$)	1.64	1.96	2.24	2.81

Table 3.3: Critical values z_α for various significance levels α of the standard normal distribution

Testing with Two Samples - Parametric

In contrast, in testing with two samples, random samples are taken from two populations, denoted as X_1 and X_2. The statistical measure of X_1 is compared with the one of X_2, e.g., comparing the arithmetic mean of X_1 with the one of X_2. To illustrate the application for OPS consider that statement: On average, the exponential growth rate of one ALG ($\mu^{(1)}$) is greater than the exponential growth rate of a benchmark algorithm ($\mu^{(2)}$), i.e., $\hat{\mathbf{E}}[\mu^{(1)}] > \hat{\mathbf{E}}[\mu^{(2)}]$.

In line with testing with one sample three hypothesis systems can be formulated. When verifying $\hat{\mathbf{E}}[X_1] \neq \hat{\mathbf{E}}[X_2]$ it is hypothesis testing with a both-sided critical region and formulated by the system

$$H^{both-sided} = \begin{cases} H_0: & \hat{\mathbf{E}}[X_1] = \hat{\mathbf{E}}[X_2] \\ H_1: & \hat{\mathbf{E}}[X_1] \neq \hat{\mathbf{E}}[X_2] \end{cases}, \tag{3.47}$$

but verifying $\hat{\mathbf{E}}[X_1] < \hat{\mathbf{E}}[X_2]$ is formulated by

$$H^{left-sided} = \begin{cases} H_0: & \hat{\mathbf{E}}[X_1] \geq \hat{\mathbf{E}}[X_2] \\ H_1: & \hat{\mathbf{E}}[X_1] < \hat{\mathbf{E}}[X_2] \end{cases} \tag{3.48}$$

and verifying $\hat{\mathbf{E}}[X_1] > \hat{\mathbf{E}}[X_2]$ is expressed by

$$H^{right-sided} = \begin{cases} H_0: & \hat{\mathbf{E}}[X_1] \leq \hat{\mathbf{E}}[X_2] \\ H_1: & \hat{\mathbf{E}}[X_1] > \hat{\mathbf{E}}[X_2] \end{cases}. \tag{3.49}$$

In the situation that X_1 and X_2 are independent and that the underling populations of them are normal distributed, then the hypothesis systems are solvable through the *two sample z-test* (TZT), or *two sample t-test* (TTT) if N is small. The test statistic for the TZT is

$$\mathcal{T}^{TZT} = \frac{\hat{\mathbf{E}}[X_1] - \hat{\mathbf{E}}[X_2]}{\sqrt{\frac{(\mathbf{SD}[X_1])^2}{N_1} + \frac{(\mathbf{SD}[X_2])^2}{N_2}}} \tag{3.50}$$

with N_1 and N_2 as the number of observations for X_1 and X_2 (The test statistic for TTT is not considered here because it is almost similar to TZT for large N_1 and N_2). The value of $\mathcal{T}^{TZT}$ is compared with a critical value of the standard normal distribution for a given significance level α (see Table 3.3). If the two samples are sampled pairwise, then it is beneficial to consider the difference of the two samples as an own random variable, i.e., $D^{(j)} = X_1^{(j)} - X_2^{(j)}$. Statements concerning $D^{(j)}$ can be treated as testing with one sample, i.e., *pairwise two sample z-test* (PTZT). The past explanations for parametric testing with two samples are made with reference to [Ott and Longnecker, 2008, pp. 290-359].

Testing with Two Samples - Nonparametric

In line with [Ott and Longnecker, 2008, pp. 290-359], the assumption for the normal distribution of data is often problematic. In this case it is useful to apply nonparametric tests as an alternative. They do not require any assumptions concerning the distribution of the underlying population of the samples. A common alternative for the TZT is the *two sample Wilcoxon signed-rank test* (TWSRT). In the TWSRT the presented hypothesis systems of Equations (3.47) – (3.49) are modified such that the relationship between the medians of the two samples are verified, instead of the arithmetic mean (i.e., the 0.5-quantile: $\hat{\mathbf{Q}}_{0.5}[X]$). But, before one can calculate the test statistic for the TWSRT some preparations are required:

(i) **Sort observations**: Sort all observations of X_1 and X_2 in one combined list from smallest to largest.

(ii) **Create rank numbers**: Give each observation a rank within the combined list. The lowest value of observation receives the rank 1 and the largest the rank $(N_1 + N_2)$.

(iii) **Handle duplicates**: If one value of observation occurs more than once in the combined list, then assign to each of those observations the average rank of the corresponding duplicates.

(iv) **Sum up ranks**: Sum up the ranks in the combined list which are originally associated with X_1. Let $SUMRANK(X_1)$ be this sum.

Thus, if the size of the two samples is sufficiently large ($N_1, N_2 > 10$), the test statistic for the TWSRT is

$$\mathcal{T}^{TWSRT} = \frac{SUMRANK(X_1) - \frac{N_1(N_1+N_2+1)}{2}}{\sqrt{\frac{N_1 N_2(N_1+N_2+1)}{12}}}. \tag{3.51}$$

and must be compared with the critical value z_α of the standard normal distribution for a given α (see Table 3.3).

If the two samples are sampled pairwise, then the *pairwise two sample Wilcoxon signed-rank test* (PTWSRT) is an alternative for the PTZT. The only condition is that the population of the difference between the pairwise observations is distributed symmetrically around the unknown median. The preparation to calculate the test statistic of the PTWSRT is as follows:

(i) **Calculate differences**: Calculate the differences $D^{(j)} = X_1^{(j)} - X_2^{(j)}$ for $j = 1, ..., N$.

(ii) **Sort differences**: Build the absolute differences and sort those from the smallest to largest.

(iii) **Delete zero values**: Remove all differences with zero value and let N' be the number of nonzero differences with $N' \leq N$.

(iv) **Create rank numbers**: Give each non-zero difference a rank. The lowest absolute difference receives the rank 1 and the largest the rank N'.

(v) **Handle duplicates**: If one value of absolute difference occurs more than once, then assign to each of those differences the average rank of the corresponding duplicates.

(vi) **Sum up ranks**: The summing up of the ranks depends on the applied hypothesis system. Let $SUMRANK(D)$ be a sum function of specific rankings. If a both-sided hypothesis system is solved, then $SUMRANK(D)$ equals the smaller rank sum of the on the one side positive differences and on the other side negative differences. If it is a left-sided (right-sided) hypothesis system, then $SUMRANK(D)$ equals the sum of ranks for the negative (positive) differences.

Finally, when $N' > 50$ then calculate the test statistic for the PTWSRT by

$$\mathcal{T}^{PTWSRT} = \frac{SUMRANK(D) - \frac{N'(N'+1)}{4}}{\sqrt{\frac{N'(N'+1)(2N'+1)}{24}}}. \tag{3.52}$$

and compare it with the critical value z_α of the standard normal distribution (see Table 3.3) for a given significance level α, but use $\mathcal{T}^{PTWSRT} \geq z_\alpha$ for $H^{left-sided}$. (Note, PTWSRT reduces the two sample data into one sample data. Thus, technically a nonparametric one sample signed-ranked test is executed.)

3.4.3 Selected Sampling Techniques

Before using an ALG in practice its performance should be tested. With reference to Black [1971, p. 19] such test should follow some rules. First, the ALG should be tested on past data. This procedure is often called historical simulation or backtesting. If this past data is used to develop the ALG, then the same data cannot be used to test the ALG. Second, the ALG should be tested on a lot of data. This means a lot of assets and even more important a lot of years should be used. Third, if it is possible, then use the ALG on future data, what means that the ALG is used to execute "fictive" trades. Fourth, make statistical tests with a consistent test design to clarify whether the good performance of an ALG is just generated due to luck.

However, if not enough data are available, (i.e., if T or m of the dataset is not large), then the performance evaluation and conclusions are problematic. To handle that problem, next to backtesting, the test of an ALG can be done by the support of sampling techniques. The idea is to generate a large number of artificial market instances by the support of permutations, leaving out and randomness. The ALG is run on each artificial market instance and the statistics of interest are calculated [according to Efron, 1979; Efron and Gong, 1983]. Selected sampling techniques are described.

Jackknife

Quenouille [1949] and Tukey [1958] propose to leave d elements out of an initial series of elements. This technique is called *jackknife approach* (JK) and can be used to estimate statistics of interest if the number of elements in the initial series is not sufficiently large enough. This means for portfolio selection. When considering variation of time then d is a number of trading days with $d < T$. Then, each sample $\mathbf{x}^{JK}$ consists of a sequence $\mathbf{x} = \mathbf{x}_1, ..., \mathbf{x}_{(T-d)}$. Theoretically $\binom{T}{d}$ different samples can be generated. If the number of assets is varied, then d is a number of assets with $d < m$. Thus, each $\mathbf{x}_t$ in $\mathbf{x}^{JK}$ consists of a vector $\left(x_{1t}, \cdots, x_{(m-d)t}\right)$. Note that a simultaneous variation of time and assets is also possible.

Bootstrap

A more widely applicable approach than the JK is the *bootstrap approach* (BT), which is introduced by Efron [1979]. Assume the observed market sequence $\mathbf{x} = \mathbf{x}_1, ..., \mathbf{x}_T$ follows a sample distribution $\hat{\mathbf{F}}$ such that $\mathbf{x} \sim \hat{\mathbf{F}}$. A BT sample $\mathbf{x}^{BT}$ *with* permutation is generated by replacing at least two elements of $\mathbf{x}$. For example, let the observed market sequence with $T = 5$ be $\mathbf{x} = \mathbf{x}_1, \mathbf{x}_2, \mathbf{x}_3, \mathbf{x}_4, \mathbf{x}_5$. Then, the BT with replacement could be $\mathbf{x}^{BT} = \mathbf{x}_1, \mathbf{x}_4, \mathbf{x}_3, \mathbf{x}_2, \mathbf{x}_5$ after one permutation. The permutation can be repeated as often as desired. In contrast, a bootstrap sample *without* replacement either reduces the number of elements in $\mathbf{x}^{BT}$ or there are repeating elements. For example, the random sample could consists of $\mathbf{x}^{BT} = \mathbf{x}_3, \mathbf{x}_1, \mathbf{x}_2$ where $T = 3$ or with repeating elements $\mathbf{x}^{BT} = \mathbf{x}_3, \mathbf{x}_3, \mathbf{x}_1, \mathbf{x}_2, \mathbf{x}_5$ where it remains $T = 5$. The advantage of the BT is, it can be shown that the joint sample distribution of all generated BT samples ($\mathbf{x}^{BT}$) follow the sample distribution $\hat{\mathbf{F}}$. The application of the BT to generate variations of assets is similar, but only meaningful *without* replacement. Otherwise, the BT sample is just the same market but with a change in the order of the assets, which has no effect on the performance of an ALG.

Cross-Validation

A simple approach to generate market sequences is to separate the initial dataset into two parts [Efron and Gong, 1983, p. 37]. One is the training part and the other one the test part. This technique is mainly used for cross-validation what is a test design to evaluate the predictive power of an arbitrary ALG, i.e., *cross-validation approach* (CV). Consider $\mathbf{x}$, delete all prices after T' with $0 < T' < T$. Then, calculate ALG on the training part with $t = 0, ..., T'$ and the test part $t = (T' + 1), ..., T$. Iterate for various T' and repeat the calculations. Compare the results of the ALG of the training part with the results of the test part. A survey of CV procedures is given by Arlot and Celisse [2010].

Random Walk

The previously described sampling techniques to generate hypothetical market instances are technically based on real observed data, i.e., nonparametric sampling techniques. However, there are approaches generating complete artificial data, i.e., parametric sampling techniques. These are based on parametric models and simulate the price movements of assets, i.e., the asset return factor for one trading period. The advantage of a model based sampling technique is that a theoretically unlimited number of hypothetical markets with arbitrary length and statistically identical characteristics can be generated, i.e., random walk approach. Note that these statistical characteristics are the input parameters of the parametric model. A statistical analysis of asset returns to verify the suitability of such models can be found in [Cont, 2001].

One model to describe price movements of assets is the *random walk model* (RW). It is assumed that the price changes are random and normal distributed. Each x_{it} is independently generated without depending on $x_{i(t-1)}$ [for further explanation of the RW in financial markets it is recommended to read Malkiel, 1973]. Let μ_i be a given exponential growth rate for A_i and σ_i the corresponding standard deviation (use Equations (3.8) and (3.15) as given values for the performance when buy and hold one single asset on an historical dataset). Then, modifying the explanations of Tsay [2010, p. 72-74] in the way that a random walk can be generated. For this, the μ_i quantifies the time trend of the RW and the term $\sigma_i X_{it}$ is the noise term with X_{it} as a standard normal distributed random variable, i.e., $X_{it} \sim \mathcal{N}(0,1)$. When using Tsay [2010, p. 72, Equation (2.35)] then $\ln(q_{it})$ is the logarithmic price of A_i at time instant t. It is generated by

$$\ln(q_{it}) = \mu_i + \ln(q_{(t-1)i}) + \sigma_i X_{it} \tag{3.53}$$

and the change of the logarithmic price can be written as

$$\ln(q_{it}) - \ln(q_{(t-1)i}) = \mu_i + \sigma_i X_{it} \tag{3.54}$$

where the left side is equivalent to $\ln\left(\frac{q_{it}}{q_{i(t-1)}}\right)$.

It is easy to see, a sequence of random return factors must be generated by

$$x_{it} = \exp(\mu_i + \sigma_i X_{it}) \tag{3.55}$$

such that the resulting asset prices follow a random walk.

To generate asset prices depending on one or more previous returns the random walk must be extended. The resulting model is called *auto-regressive model* (AR). As the RW, in the AR μ_i and σ_i must be given plus an α which quantifies the impact of the return from the previous trading period on the current trading period. It is provided a formula which takes into account only one previous trading period. Thus, in accordance with [Tsay, 2010, p.

75] a sequence of return factors with the AR for $t = 1, ..., T$ is generated by

$$x_{it}^{AR} = \exp\left(\mu_i + \sigma_i X_{it} + \alpha \ln(x_{i(t-1)})\right) \tag{3.56}$$

where x_{i0} can be set to 1.

In contrast, if x_{it} depends not on the full $x_{i(t-1)}$ but only on the previous random variable $X_{i(t-1)}$, then this is called *moving average model* (MA) [on the basis of Tsay, 2010, pp. 64-71]. Thus, a sequence of return factors with the MA for $t = 1, ..., T$ is generated by

$$x_{it}^{MA} = \exp\left(\mu_i + \sigma_i X_{it} + \beta \sigma_i X_{i(t-1)}\right) \tag{3.57}$$

where β must be a given impact factor. When AR and MA are combined, the resulting *auto-regressive moving average model* (ARMA) is defined as

$$x_{it}^{ARMA} = \exp\left(\mu_i + \sigma_i X_{it} + \alpha \ln(x_{i(t-1)}) + \beta \sigma_i X_{i(t-1)}\right). \tag{3.58}$$

Note that the models can be extended such that they take more subsequent trading periods and/or correlation between asset returns into account. The application of sampling techniques which generate artificial data to evaluate the performance of an ALG can be found for example in Brock et al. [1992] and DeMiguel et al. [2009].

3.5 Competitive Analysis

Competitive analysis is a concept to analyze an online ALG which must satisfy an unpredictable sequence of requests. The performance of that online ALG is compared to the performance of an offline ALG (OPT) that knows the entire input sequence in advance. For the PSP the input consists of the market $\mathbf{x}$ containing m assets and T trading periods. Let $Perf(\cdot)$ be a function for an arbitrary performance measure as described in Section 3.2.

From a game-theoretic perspective, the online ALG is executed by an online player and is compared to the offline player as an omniscient adversary that runs OPT [see also Albers, 2006, p. 143]. According to Koutsoupias and Papadimitriou [2000, pp. 300-303] the evaluation of three types of the worst-case competitiveness of an online ALG are useful, i.e., (i) competitive ratio, (ii) performance ratio and (iii) comparative ratio. The ratio is the relation between the output of the online player and the offline player. In addition, Fujiwara et al. [2011] propose the average-case competitive analysis to analyze an ALG in an average-case scenario. Thus, the (iv) average-case competitive ratio is provided.

3.5.1 Competitive Ratio

Since the PSP is a maximization[2] problem the *competitive ratio* of an online ALG depending on the input sequence $\mathbf{x}$ has been defined as

$$c = \max_{\mathbf{x}} \frac{Perf(OPT, \mathbf{x})}{Perf(ALG, \mathbf{x})} \tag{3.59}$$

where $\mathbf{x}$ ranges over all possible markets [compare with Koutsoupias and Papadimitriou, 2000, p. 300]. It is easy to see that $c \geq 1$ and that the smaller c the more powerful is the online ALG. It follows, an ALG is c-competitive if the lower bound for the performance can be formulated as

$$Perf(ALG, \mathbf{x}) \geq \frac{1}{c} Perf(OPT, \mathbf{x}), \tag{3.60}$$

which must be valid for any market $\mathbf{x}$ [see also Manasse et al., 1988, pp. 322, 323]. Note that there are no statistical assumptions formulated for $\mathbf{x}$.

3.5.2 Performance Ratio

When adding the assumption that the actual instance of the market $\mathbf{x}$ is from a class $\mathbf{X}$ containing all possible market instances which meet one or more specific properties, then the *performance ratio*

$$c(\mathbf{X}) = \max_{\mathbf{x} \in \mathbf{X}} \frac{Perf(OPT, \mathbf{x})}{Perf(ALG, \mathbf{x})} \tag{3.61}$$

is a refinement of the competitive ratio [see also Koutsoupias and Papadimitriou, 2000, p. 301]. If $\mathbf{X}$ consists of all possible instances of the market $\mathbf{x}$ and the one with the worst-case scenario is assigned to a probability of one and zero to the others, then $c(\mathbf{X}) = c$. Thus, the adversary picks the instance with the best performance and the online player the one with the worst. It is obvious, Equation (3.60) changes to $Perf(ALG, \mathbf{x}) \geq \frac{1}{c(\mathbf{X})} Perf(OPT, \mathbf{x})$ and is only valid for $\mathbf{x} \in \mathbf{X}$.

[2] In a minimization problem the competitive ratio would be defined as $c = \max_{\mathbf{x}} \frac{Perf(ALG, \mathbf{x})}{Perf(OPT, \mathbf{x})}$

3.5.3 Comparative Ratio

Assume the adversary has to choose one algorithm B from a set of algorithms $\mathfrak{B}$, such that $B \in \mathfrak{B}$. Due to the restriction to a set of algorithms $\mathfrak{B}$ the power of the adversary is restricted. Note that the online ALG may also be included in $\mathfrak{B}$. From this perspective the *comparative ratio* is formulated as

$$c(ALG, \mathfrak{B}) = \max_{B \in \mathfrak{B}} \max_{\mathbf{x}} \frac{Perf(B, \mathbf{x})}{Perf(ALG, \mathbf{x})} \tag{3.62}$$

what is one more refinement of the origin competitive ratio given in Equation (3.59) [see Koutsoupias and Papadimitriou, 2000, p. 302]. Note, if $\mathfrak{B}$ consists of all possible online and offline algorithms, then $c(ALG, \mathfrak{B}) = c$. For the sake of completeness, combining the ideas of Equation (3.61) and (3.62) it follows the *comparative performance ratio*

$$c(ALG, \mathfrak{B}, \mathbf{X}) = \max_{B \in \mathfrak{B}} \max_{\mathbf{x} \in \mathbf{X}} \frac{Perf(B, \mathbf{x})}{Perf(ALG, \mathbf{x})} \tag{3.63}$$

which compares the ALG under the assumption of limited choices for the adversary concerning (i) the algorithms and (ii) the class of market instances. It is clear, if $\mathfrak{B}$ consists of all possible online and offline algorithms and $\mathbf{X}$ contains all possible market instances, then $c(ALG, \mathfrak{B}, \mathbf{X}) = c(ALG, \mathfrak{B}) = c(\mathbf{X}) = c$.

3.5.4 Average-Case Competitive Ratio

With reference to Fujiwara et al. [2011, p. 86], the performance evaluation of an online ALG with a worst-case competitive analysis seems to be often too pessimistic. The performance of an online ALG with the best worst-case competitive ratio can be in practice on "normal" instances far away from optimal. In addition, sometimes the worst-case competitive ratio cannot tell the difference between the algorithms properly. This means, algorithms with equal competitive ratio may perform different on "normal" instances. These drawbacks are encountered by the average-case competitive analysis. An ALG has an *average-case performance ratio* of

$$\mathrm{E}\left[c(\mathbf{X})\right] = \mathop{\mathrm{E}}_{\mathbf{x} \in \mathbf{X}} \left[\frac{Perf(OPT, \mathbf{x})}{Perf(ALG, \mathbf{x})}\right] \tag{3.64}$$

under the assumption that all $\mathbf{x}$ are from a class of possible inputs $\mathbf{X}$, where $\mathbf{X}$ is based on one given statistical distribution. Note, this is different to worst-case competitive analysis. An *average-case comparative ratio* would be defined in the same way. Fujiwara et al. [2011] discuss an alternative average-case ratio $\frac{\mathrm{E}[OPT(\mathbf{x})]}{\mathrm{E}[ALG(\mathbf{x})]}$ and notify that results may differ depending on which measure is adopted.

According to Fujiwara et al. [see 2011, p. 106], a drawback of this measure is, the measure is only analytically determinable if the input $\mathbf{X}$ has a simple structure and is explicitly parameterized. In the financial context, it is obvious to assume as a first step that all instances of $\mathbf{X}$ follow a geometric Brownian motion.

Note that Spielman and Teng [2004] proposed a hybrid of worst-case and average-case analysis to evaluate the performance of an ALG, i.e., smoothed analysis. Nevertheless, till now average-case analysis and smoothed analysis for the PSP are not studied sufficiently.

3.5.5 Concept of Universality

According to Cover [1991], an online ALG that solves the PSP is *universal* when it satisfies

$$\frac{1}{T}\ln Perf(ALG,\mathbf{x}) - \frac{1}{T}\ln \max_{B\in\mathfrak{B}} Perf(B,\mathbf{x}) \longrightarrow 0 \tag{3.65}$$

for an increasing investment horizon T with $Perf(\cdot) = W_T$ and $\mathfrak{B}$ as a set of constant rebalancing algorithms. Obviously, the exponential growth rate of the ALG has to converge asymptotically to the exponential growth rate of the BCR for increasing T. Equation (3.65) has to be true for any market sequence $\mathbf{x}$, following Borodin et al. [2004, p. 581]. The concept of universality can be limited to a worst-case scenario [see Borodin et al., 2000, p. 175]. It can be shown that any ALG is universal if and only if the comparative ratio is subexponential (indeed polynomial) [see Borodin et al., 2004, p. 581]. Because

$$\frac{1}{T}\left(\ln W_T(ALG,\mathbf{x}) - \ln \max_{B\in\mathfrak{B}} W_T(B,\mathbf{x})\right) \longrightarrow 0 \tag{3.66}$$

and this is equal to

$$\frac{1}{T}\ln\left(\frac{W_T(ALG,\mathbf{x})}{\max_{B\in\mathfrak{B}} W_T(B,\mathbf{x})}\right) \longrightarrow 0 \tag{3.67}$$

[see Gaivoronski and Stella, 2000, p. 176] with $W_T(\cdot)$ as a function for the terminal wealth depending on a given ALG and market $\mathbf{x}$. By replacing the ratio through the comparative ratio it can be seen that this is equal to

$$\frac{1}{T}\ln\frac{1}{c(ALG,\mathfrak{B})} \longrightarrow 0, \tag{3.68}$$

which quantifies the *extent of universality*. In Table 3.4 examples are given to show how the extent of universality as shown in Equation (3.68) reacts for fictive types of the comparative ratio with different investment horizons. When the comparative ratio is exponential ($c(ALG,\mathfrak{B}) = 1.02^T$), the extent of universality is not converging to zero for increasing T. In contrast, it does for a logarithmic ($c(ALG,\mathfrak{B}) = \ln T$) and constant ($c(ALG,\mathfrak{B}) = 5$) comparative ratio.

$c(ALG,\mathfrak{B})$	$T=1$	$T=10$	$T=100$	$T=200$	$T=300$
1.02^T	-0.020	-0.020	-0.020	-0.020	-0.020
$\ln T$	-	-0.083	-0.015	-0.008	-0.006
5	-1.609	-0.161	-0.016	-0.008	-0.005

Table 3.4: Extent of universality for fictive comparative ratios and increasing investment horizon T

In this context, the performance of an online ALG that solves the PSP is often expressed by the *regret*, what is defined as

$$regret = -\ln\frac{1}{c(ALG,\mathfrak{B})} = \ln c(ALG,\mathfrak{B}) \tag{3.69}$$

such that $c(ALG,\mathfrak{B}) = \exp(regret)$ [refer to Li and Hoi, 2014, p. 8]. It is easy to see, an investor who tracks the BCR aims to achieve a low as possible regret.

Now, the worst-case scenario for the PSP is discussed. Cover and Ordentlich [1996, p. 350] and Borodin et al. [2000, p. 177] describe such a scenario as a Kelly sequence[3] where the market $\mathbf{x}$ consists only of one winner and $(m-1)$ loser asset(s). Wealth invested in the winner asset remains the same for each trading period. But, the wealth invested in the loser asset(s) always decreases ($x_{it} < 1$), where the maximum loss is

[3] A Kelly sequence is based on the ideas of Kelly gambling. It is referred to [Kelly, 1956].

a total loss ($x_{it} = 0$), i.e., a crash of the asset. Thus, an example market following a Kelly sequence with $m = 6$ and maximum loss would be

$$\mathbf{x}_t = (0, 1, 0, 0, 0, 0) \tag{3.70}$$

where, except A_2 as the winner asset, all assets crash.

3.5.6 Competitive Ratio as Performance Measure

With reference to [al-Binali, 1999] and [Ahmad et al., 2014], the competitive ratio (plus performance ratio and comparative ratio) itself is also a performance measure. Note, from a simplified perspective, a performance measure can be any function that quantifies for a specific market instance the return on investment, risk or risk-adjusted performance of an ALG. Consider the execution of a competitive analysis for an online ALG as an operation with three components, where each component consists of, at least, one particular performance measure:

(i) **Calculating ALG**: During the investment horizon, assume the calculation of ALG is based on a periodical optimization. The objective function of that optimization can consists of a specific performance measure $Perf_1(ALG, \mathbf{x}_1, ..., \mathbf{x}_t)$, which is quantified by information up to trading period t. (For examples of algorithms, which take performance measures into account consider the capital growth form in Table 4.3 on Page 107, where basically performance measures for return on investment are employed. In addition, the proposed algorithms of Chapter 5 incorporate a risk measure.)

(ii) **Measuring Performance**: When ALG is calculated for the entire investment horizon $t = 1, ..., T$, then an additional performance measure $Perf_2(ALG, \mathbf{x})$ is required, which quantifies the performance of ALG for a given $\mathbf{x}$. Note that $Perf_1(ALG, \mathbf{x}_1, ..., \mathbf{x}_t)$ could be a performance measure for return on investment and $Perf_2(ALG, \mathbf{x})$ for return on investment, risk or risk-adjusted performance or vice versa.

(iii) **Determining Competitive Ratio**: A statement concerning the competitiveness of ALG is always based on a given $Perf_2(\cdot)$. Independently on the concrete $\mathbf{x}$, a maximum value for the relative distance between $Perf_2(ALG, \mathbf{x})$ and $Perf_2(OPT, \mathbf{x})$ can be quantified, i.e.,

$$Perf_3(ALG, \mathbf{x}) = \max_{\mathbf{x}} \frac{Perf_2(OPT, \mathbf{x})}{Perf_2(ALG, \mathbf{x})}, \tag{3.71}$$

which is equal to Equation (3.59). Ahmad et al. [2014] define $Perf_3(\cdot)$ as a measure of risk. Hence, $Perf_3(ALG, \mathbf{x})$ quantifies the risk of $Perf_2(ALG, \mathbf{x})$ through the consideration of a worst-case scenario.

Summing up, one way to interpret the three different performance measures during the execution of a competitive analysis is as follows. $Perf_1(\cdot)$ and $Perf_2(\cdot)$ are usually determined based on a concrete input sequence $\mathbf{x}$ and are therefore deemed to be only for the empirical performance evaluation of an ALG. $Perf_3(\cdot)$ can only be determined analytically. Its value is universally valid and is independent of a concrete $\mathbf{x}$. In other words, $Perf_1(\cdot)$ and $Perf_2(\cdot)$ are empirical performance measures and $Perf_3(\cdot)$ is an analytical performance measure. However, all three support the evaluation of a given online ALG.

3.6 Conclusions

In this chapter, the main performance measures and selected benchmarks to evaluate the performance of an online ALG for the PSP are provided. In addition, to make a statement for the performance of an ALG with respect to the uncertainty of asset prices in the market the two concepts statistical analysis and competitive analysis are outlined. For the statistical analysis selected statistical measures, hypothesis testing and sampling techniques are summarized. For the competitive analysis various types to determine a performance guarantee for an online ALG based on worst-case and average-case scenarios are provided. Summing up, this chapter facilitates the understanding of the different perspectives of the FC and the MLC on possible solutions for the PSP.

4 Selected Algorithms from the Literature

This chapter surveys a selection of online algorithms concerning the portfolio selection problem. The selection is motivated by the survey of Li and Hoi [2014]. In this chapter, the presentation of algorithms differs in the form that the algorithms are described from a technical perspective with the support of pseudocodes. This allows a better understanding of what the algorithms exactly do at one glance and simplifies the implementation in a software tool.

4.1 Preliminaries

Recall the following PSP. An investor with an investment horizon of $t = 1, ..., T$ trading periods has to decide at the beginning of each trading period how to allocate the current wealth on $A_1, ..., A_m$ assets. Let

$$\mathbf{x}_t = (x_{1t}, \ldots, x_{mt}) \tag{4.1}$$

be a vector presenting the return factors of the $i = 1, ..., m$ assets during trading period t with x_{it} as the return factor of A_i during trading period t. The sequence $\mathbf{x} = \mathbf{x}_1, \ldots, \mathbf{x}_T$ represents the entire market. At the beginning of each trading period an allocation of wealth

$$\mathbf{b}_t = (b_{1t}, \ldots, b_{mt}) \tag{4.2}$$

is requested with b_{it} as a proportion of wealth. The sequence of decisions is denoted as $\mathbf{b} = \mathbf{b}_1, \ldots, \mathbf{b}_T$.

The problem is, how the investor should choose each $\mathbf{b}_t$ in order to maximize

$$W_T = W_0 \prod_{t=1}^{T} \sum_{i=1}^{m} x_{it} b_{it} \tag{4.3}$$

with respect to the uncertainty of the market $\mathbf{x}$. According to [Borodin et al., 2004] this problem can be solved by an online ALG. The survey of Li and Hoi [2014] summarizes and classifies prominent online algorithms from the literature of the MLC. This chapter limits to the types of algorithms *Follow-the-Winner* (FTW) and *Follow-the-Loser* (FTL) of this survey. FTW algorithms are characterized by increasing proportions in the assets (or experts) which performed well in the past. Algorithms in this category target the same exponential growth rate as the BCR. Thus, any ALG from the type FTW is universal (for the concept of universality see Section 3.5.5). Algorithms from the type FTL overweight the loser assets of the past. The underlying idea behind buying losers is anti-cyclical investment behavior to exploit mean reversion, which means that the asset prices fluctuate around a mean value. Building portfolios by buying losers and selling winners in stock markets is also discussed in the FC, e.g., [DeBondt and Thaler, 1985], [Jegadeesh and Titman, 1993], [Richards, 1997] and [Malin and Bornholt, 2013].

In this chapter, each online ALG from the type FTW and FTL discussed in the survey of Li and Hoi [2014] is presented from a technical perspective with the support of a pseudocode. This gives a better understanding of each ALG and facilitates its implementation in a software tool for scientists and practitioners. Before the algorithms are provided, the underlying ideas of the types FTW and FTL algorithms are shown exemplary in a virtual market.

4.1.1 Virtual Market

Referring to Cover and Gluss [1986], consider a sequence of

$$\mathbf{x}_t = \begin{cases} (1 \quad 2) & \text{if } t \text{ is odd} \\ (1 \quad \frac{1}{2}) & \text{if } t \text{ is even,} \end{cases} \tag{4.4}$$

for $t = 1, ..., T$ as a virtual market $\mathbf{x} = \mathbf{x}_1, ..., \mathbf{x}_T$ with $m = 2$. The A_1 is a cash asset with no change in price. The investment in A_1 is risk-free and without any growth. The A_2 is a highly volatile asset. Its value doubles on odd trading periods and halves on even trading periods. The aim is to maximize the terminal wealth W_T, respectively the exponential growth rate of wealth μ (see Section 3.2.1). The investor does not know whether A_1 or A_2 is the cash asset. However, the investor wants to earn a $\mu > 0$ in any case and has to choose one of three investment strategies, i.e., (i) *buy-and-hold portfolio* (BH), (ii) *constant rebalancing portfolio* (CR) and (iii) *switching portfolio* (SW).

Some preliminaries - the decision problem consists of two parts. One part is to choose one of the three investment strategies. The second one is, to commit to only one portfolio allocation $\mathfrak{b}$. The investor cannot predict the future development of the virtual market. Short-selling and consumption of wealth is excluded. During the investment horizon the investor cannot change the investment strategy or the specified portfolio allocation $\mathfrak{b}$. To be more precise, $\mathfrak{b}$ must be chosen from the simplex

$$\mathfrak{B}_m = \left(\mathfrak{b} \in \mathbb{R}^m : \mathfrak{b}_i \geq 0, \sum_{i=1}^{m} \mathfrak{b}_i = 1 \right) \tag{4.5}$$

with $m = 2$ and combined with one of the investment strategies:

(i) **Buy-And-Hold Portfolio (BH)**: At the beginning of trading period $t = 1$ the portfolio allocation $\mathfrak{b}$ is realized once. In the subsequent trading periods, no adjustment of the portfolio allocation takes place. Thus, the terminal wealth is calculated by

$$W_T^{BH}(\mathfrak{b}) = W_0 \sum_{i=1}^{m} \mathfrak{b}_i \prod_{t=1}^{T} x_{it} \tag{4.6}$$

with $m = 2$.

(ii) **Constant Rebalancing Portfolio (CR)**: In $t = 1$ the wealth is initially allocated by the proportions of $\mathfrak{b}$ and held constant for all $t = 2, ..., T$ trading periods, i.e., constant rebalancing. The terminal wealth is determined by

$$W_T^{CR}(\mathfrak{b}) = W_0 \prod_{t=1}^{T} \sum_{i=1}^{m} x_{it} \mathfrak{b}_i \tag{4.7}$$

with $m = 2$. This is equal to Equation (4.3) with $\mathbf{b}_{(t+1)} = \mathbf{b}_t$ for $t = 1, ..., (T-1)$.

(iii) **Switching Portfolio (SW)**: The switching portfolio strategy in this form cannot be found directly in the current literature but here it is motivated by Singer [1997] and Li and Hoi [2014, Example 3.2] to explain the FTL concept. In odd trading periods the wealth is allocated by the portfolio allocation $\mathbf{b}_t = (\mathfrak{b}_1, \mathfrak{b}_2)$ and in even trading periods the portfolio allocation switches by $\mathbf{b}_t = (1 - \mathfrak{b}_1, 1 - \mathfrak{b}_2)$. The terminal wealth for the switching portfolio strategy is calculated by

$$W_T^{SW}(\mathfrak{b}) = W_0 \prod_{t=1}^{T} \sum_{i=1}^{m} x_{it} \left((t \bmod 2)\mathfrak{b}_i + ((t-1) \bmod 2)(1 - \mathfrak{b}_i)\right) \quad (4.8)$$

with $m = 2$ and includes two modulo operations.

The two-part decision problem can be solved when considering the terminal wealth W_T, respectively the exponential growth rate μ of each investment strategy for arbitrary $\mathfrak{b}$. For simplification, assume in the following $W_0 = 1$. Since $\mathfrak{b}$ with $m = 2$ can be reduced to $\mathfrak{b} = (\mathfrak{b}_1, 1 - \mathfrak{b}_1)$ the terminal wealth for the three investment strategies has to be calculated depending only on $\mathfrak{b}_1$.

Calculations to Buy-and-Hold Portfolio Strategy

The terminal wealth of the BH for arbitrary T is calculated by including the virtual market given by Equation (4.4) into the Equation (4.6). Thus, it is concrete

$$W_T^{BH}(\mathfrak{b}) = \begin{cases} \mathfrak{b}_1 1^{\frac{T+1}{2}} 1^{\frac{T-1}{2}} + (1 - \mathfrak{b}_1) 2^{\frac{T+1}{2}} \frac{1}{2}^{\frac{T-1}{2}} & \text{if } T \text{ is odd} \\ \mathfrak{b}_1 1^{\frac{T}{2}} 1^{\frac{T}{2}} + (1 - \mathfrak{b}_1) 2^{\frac{T}{2}} \frac{1}{2}^{\frac{T}{2}} & \text{if } T \text{ is even,} \end{cases} \quad (4.9)$$

which can be shortened to

$$W_T^{BH}(\mathfrak{b}) = \begin{cases} \mathfrak{b}_1 + (1 - \mathfrak{b}_1)2 \quad = 2 - \mathfrak{b}_1 & \text{if } T \text{ is odd} \\ \mathfrak{b}_1 + (1 - \mathfrak{b}_1) \quad = 1 & \text{if } T \text{ is even.} \end{cases} \quad (4.10)$$

This results in an exponential growth rate ($\mu = \frac{1}{T} \ln W_T(\mathfrak{b})$) of

$$\mu^{BH}(\mathfrak{b}) = \begin{cases} \frac{1}{T} \ln(2 - \mathfrak{b}_1) & \text{if } T \text{ is odd} \\ 0 & \text{if } T \text{ is even,} \end{cases} \quad (4.11)$$

where it is easy to see that the limit of μ^{BH} for increasing T is

$$\lim_{T \to \infty} \mu^{BH} = 0. \quad (4.12)$$

for any allocation of wealth $\mathfrak{b}$ with $0 \leq \mathfrak{b}_1 \leq 1$. In other words, in the long run, in the virtual market it is not possible to earn any wealth with BH.

This includes also a strategy with buy and hold only one single asset, i.e., $\mathfrak{b}_1 = 0.0$ or $\mathfrak{b}_1 = 1.0$.

Calculations to Constant Rebalancing Portfolio Strategy

In contrast, the terminal wealth of the CR for any T is calculated with Equation (4.7). Precisely by

$$W_T^{CR}(\mathfrak{b}) = \begin{cases} (\mathfrak{b}_1 + (1-\mathfrak{b}_1)2)^{\frac{T+1}{2}}(\mathfrak{b}_1 + (1-\mathfrak{b}_1)\frac{1}{2})^{\frac{T-1}{2}} & \text{if } T \text{ is odd} \\ (\mathfrak{b}_1 + (1-\mathfrak{b}_1)2)^{\frac{T}{2}}(\mathfrak{b}_1 + (1-\mathfrak{b}_1)\frac{1}{2})^{\frac{T}{2}} & \text{if } T \text{ is even} \end{cases} \tag{4.13}$$

and this can be shortened to

$$W_T^{CR}(\mathfrak{b}) = \begin{cases} 2^{\frac{1}{2}-\frac{T}{2}}(2-\mathfrak{b}_1)^{\frac{T+1}{2}}(\mathfrak{b}_1+1)^{\frac{T-1}{2}} & \text{if } T \text{ is odd} \\ 2^{-\frac{T}{2}}(-(\mathfrak{b}_1)^2+\mathfrak{b}_1+2)^{\frac{T}{2}} & \text{if } T \text{ is even.} \end{cases} \tag{4.14}$$

The exponential growth rate is determined by

$$\mu^{CR}(\mathfrak{b}) = \begin{cases} \frac{1}{T}\ln\left(2^{\frac{1}{2}-\frac{T}{2}}(2-\mathfrak{b}_1)^{\frac{T+1}{2}}(\mathfrak{b}_1+1)^{\frac{T-1}{2}}\right) & \text{if } T \text{ is odd} \\ \frac{1}{T}\ln\left(2^{-\frac{T}{2}}(-(\mathfrak{b}_1)^2+\mathfrak{b}_1+2)^{\frac{T}{2}}\right) & \text{if } T \text{ is even,} \end{cases} \tag{4.15}$$

where the limit must be calculated depending on $\mathfrak{b}_1$. Table 4.1 provides the limit for various $\mathfrak{b}_1$. Noteworthy is, except for $\mathfrak{b}_1 = 0.0$ and $\mathfrak{b}_1 = 1.0$ all μ^{CR} are positive. If the investor selects a $\mathfrak{b}_1$ with $0.0 < \mathfrak{b}_1 < 1.0$ and a sufficiently large T, then the wealth invested will increase in any case. The largest μ^{CR} can be achieved with $\mathfrak{b}_1^* = 0.5$.

$\mathfrak{b}_1$	0.0	0.1	0.2	0.3	0.4	0.5
$\lim_{T\to\infty} \mu^{CR}(\mathfrak{b})$	0.000	0.022	0.038	0.050	0.057	0.059
$\mathfrak{b}_1$	-	0.6	0.7	0.8	0.9	1.0
$\lim_{T\to\infty} \mu^{CR}(\mathfrak{b})$		0.057	0.050	0.038	0.022	0.000

Table 4.1: Limit of μ^{CR} for $T \to \infty$ with various $\mathfrak{b}_1$

Calculations to Switching Portfolio Strategy

In order to find the limit of the exponential growth rate for the SW consider the terminal wealth

$$W_T^{SW}(\mathfrak{b}) = \begin{cases} (\mathfrak{b}_1 + (1-\mathfrak{b}_1)2)^{\frac{T+1}{2}}((1-\mathfrak{b}_1)+\mathfrak{b}_1\frac{1}{2})^{\frac{T-1}{2}} & \text{if } T \text{ is odd} \\ (\mathfrak{b}_1 + (1-\mathfrak{b}_1)2)^{\frac{T}{2}}((1-\mathfrak{b}_1)+\mathfrak{b}_1\frac{1}{2})^{\frac{T}{2}} & \text{if } T \text{ is even} \end{cases} \tag{4.16}$$

by using Equation (4.8). This can be shortened to

$$W_T^{SW}(\mathfrak{b}) = \begin{cases} 2^{\frac{1}{2}-\frac{T}{2}}(2-\mathfrak{b}_1)^T & \text{if } T \text{ is odd} \\ 2^{-\frac{T}{2}}(2-\mathfrak{b}_1)^T & \text{if } T \text{ is even.} \end{cases} \tag{4.17}$$

The exponential growth rate is determined by

$$\mu^{SW}(\mathfrak{b}) = \begin{cases} \frac{1}{T}\ln\left(2^{\frac{1}{2}-\frac{T}{2}}(2-\mathfrak{b}_1)^T\right) & \text{if } T \text{ is odd} \\ \frac{1}{T}\ln\left(2^{-\frac{T}{2}}(2-\mathfrak{b}_1)^T\right) & \text{if } T \text{ is even} \end{cases} \tag{4.18}$$

where the limit also depends on $\mathfrak{b}_1$. Table 4.2 provides the limit for various $\mathfrak{b}_1$. Obviously, for $\mathfrak{b}_1 \leq 0.5$ the growth rate is positive and much larger than for the CR. With $\mathfrak{b}_1 = 0.5$ the SW is exactly a copy of the CR. It can be shown that, only when $\mathfrak{b}_1 = 0.586$ the limit becomes zero, i.e., $\lim_{T\to\infty} \mu^{SW} = 0$. Thus, for $\mathfrak{b}_1 < 0.586$ the exponential growth rate is positive and invested wealth always increase with sufficiently large T. With $\mathfrak{b}_1 > 0.5$ the investor puts wealth to the volatile asset when the price is going down. However, even with the wrong "timing", the exponential growth rate remains positive until $\mathfrak{b}_1 < 0.586$. This is due to the fact that in the range $0.5 < \mathfrak{b}_1 < 0.586$ the losses caused by the wrong "timing" are compensated by the gains of the constant rebalancing.

$\mathfrak{b}_1$	0.0	0.1	0.2	0.3	0.4	0.5
$\lim_{T\to\infty} \mu^{SW}(\mathfrak{b})$	0.347	0.295	0.241	0.184	0.123	0.059
$\mathfrak{b}_1$	-	0.6	0.7	0.8	0.9	1.0
$\lim_{T\to\infty} \mu^{SW}(\mathfrak{b})$		-0.010	-0.084	-0.164	-0.251	-0.347

Table 4.2: Limit of μ^{SW} for $T \to \infty$ with various $\mathfrak{b}_1$

Results and Remarks

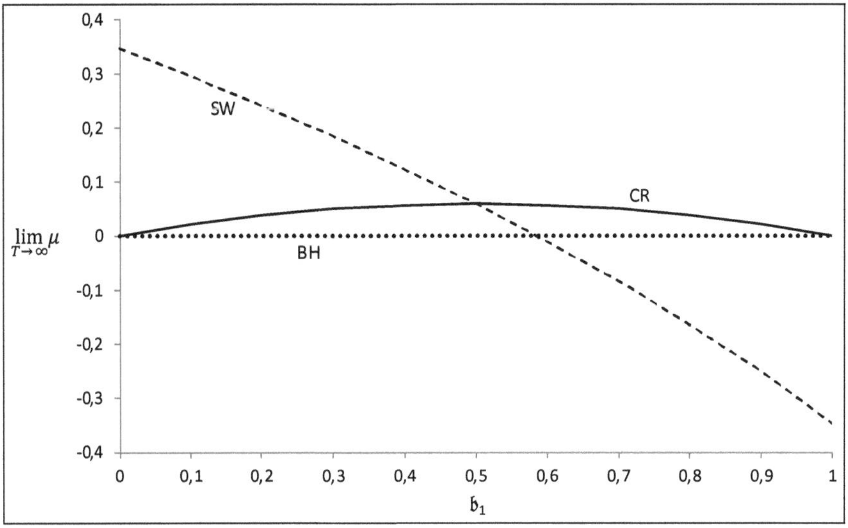

Figure 4.1: Limit of the exponential growth rate μ in the virtual market for various $\mathfrak{b}_1$: (i) BH (dotted line), (ii) CR (black line) and (iii) SW (dashed line)

To conclude, the limit of the exponential growth rate μ for BH, CR and SW are illustrated together in Figure 4.1. On the abscissa is the fraction $\mathfrak{b}_1$ and on the ordinate the limit of the exponential growth rate of wealth μ. Figure 4.2 demonstrates the achieved terminal wealth of the three strategies in the virtual market for various instances of $\mathfrak{b}_1$. One can see, in the virtual market any allocation with BH results in a zero growth of wealth. Remarkable, at the beginning of any even trading period there exists one allocation which consists only of the best asset. In the virtual market the exponential growth rate of the best asset converges to zero if T is sufficiently large. Since the investor has to achieve a $\mu > 0$ the BH strategy is not beneficial. In contrast, using the CR strategy with $0 < \mathfrak{b}_1 < 1$ the investor achieves always a $\mu > 0$, regardless of whether the cash asset is A_1 or A_2. Within the possible instances of the CR strategy, there exists one $\mathfrak{b}_1^*$ which generates the highest exponential growth rate. When considering the SW strategy it is conspicuous that the investor can achieve a much higher $\mu > 0$ than with CR and $\mathfrak{b}_1^*$. Since the investor does not know which asset is the cash asset the result of the investor can be $\mu < 0$. Thus, under the conditions

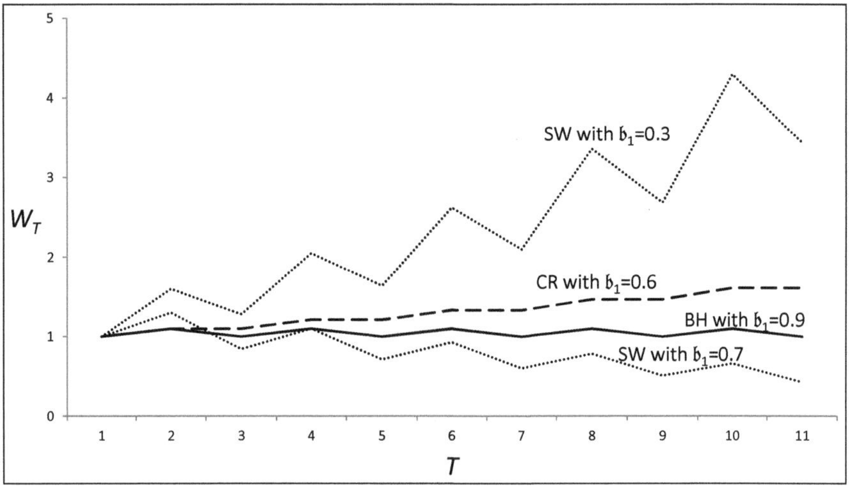

Figure 4.2: Terminal wealth W_T in the virtual market for various T and $\mathfrak{b}_1$: (i) SW with $\mathfrak{b}_1 = 0.7$ (lowest line), (ii) BH with $\mathfrak{b}_1 = 0.9$ (second lowest line), (iii) CR with $\mathfrak{b}_1 = 0.6$ (second largest line) and (iv) SW with $\mathfrak{b}_1 = 0.3$ (largest line)

described, the CR strategy with $0 < \mathfrak{b}_1 < 1$ is the only suitable strategy for the investor.

Executing the investment strategies BH, CR and SW in this virtual market allows a simplified perspective on the types of algorithms FTW and FTL. It can always be developed a learning algorithm that tries to follow a certain benchmark algorithm. This benchmark algorithm could be any fix portfolio allocation $\mathfrak{b}$ in combination with a BH, CR or SW strategy. Assume there exists for each BH, CR and SW one optimal $\mathfrak{b}^*$ as optimal benchmark algorithms. In contrast to the benchmark algorithms, an online ALG varies the portfolio allocations $\mathbf{b}_1, ..., \mathbf{b}_T$ during the investment horizon. Very simplified it can be said, if the online ALG tracks the $\mathfrak{b}^*$ of CR (achieving the same optimal exponential growth rate), then it belongs to the family of FTW algorithms. If this behavior can be analytically proofed for all possible market instance $\mathbf{x}$, then such ALG is called universal (see concept of universality in Section 3.5.5)). In the broad sense, if an online ALG tracks the $\mathfrak{b}^*$ of SW, it belongs to the type of algorithms FTL. However, tracking $\mathfrak{b}^*$ of BH is not considered because CR is superior to BH with respect to the exponential growth rate.

4.1.2 Projection onto a Simplex

An approach often used for the types of algorithms FTL is to generate an intermediate allocation $\mathbf{b}'_t = (b'_{1t}, ..., b'_{mt})$ which ignores initially the assumption of no short-selling ($b_{it} \geq 0$). At the end of the calculations of the ALG the resulting $\mathbf{b}'_t$ is normalized by a projection onto the simplex $\mathfrak{B}_m$ given in Equation (4.5) such that no short-selling is fulfilled. More precisely, the minimization problem

$$\mathbf{b}_t = \arg \min_{\mathfrak{b} \in \mathfrak{B}_m} ||\mathfrak{b} - \mathbf{b}'_t|| \tag{4.19}$$

is solved. In other words, it is asked for the $\mathfrak{b}$ with the smallest distance to $\mathbf{b}'_t$. To solve this problem Chen and Ye [2011] provide an algorithm which is presented as pseudocode in Algorithm 1. Such kind of projection algorithm is required by the considered online algorithms presented in [Li et al., 2012], [Li and Hoi, 2012], [Li et al., 2013] and [Huang et al., 2013], which are described in details in the following of this section.

4.1.3 Information and Algorithms

In the literature, there are numerous approaches to create an online ALG for the PSP. The information in a risky market $\mathbf{x}$ is processed in different ways. All have in common that past asset prices (or return factors) are used to derive a decision for the portfolio allocation at the beginning of the next trading period ($\mathbf{b}_{(t+1)}$). At the beginning of the investment horizon limited information is available, such that all online algorithms require at least one or more trading periods to produce a $\mathbf{b}_{(t+1)}$. For the required trading periods until an online ALG is able to produce $\mathbf{b}_{(t+1)}$, it is meaningful to use an allocation which is independent of information. In general, it is reasonable to assume that all online algorithms start with the naive diversification, e.g., $\mathbf{b}_1 = \left(\frac{1}{m} \cdots \frac{1}{m}\right)$ (see also Section 3.3.3).

However, a selection of how to extract information from past asset prices (or return factors) is exemplary provided.

Experts

Consider an expert as a fictive investor who participates on the market $\mathbf{x}$ with an arbitrary strategy and initial wealth $W_0 = 1$. Assume there exist $\omega = 1, ..., \Omega$ experts and each one uses its own strategy. Let the period wealth of the expert ω at the end of trading period t be denoted as W_t^{ω}. If the expert is using the specific strategy, CR strategy with fix portfolio

Algorithm 1 Projection Algorithm

```
function PROJECTION(b'_t)
    y := b'_t                                   ▷ y = y_1, ..., y_m
    Sort(y)                                     ▷ increasing order
    th := 0
    for i = m to 2 do step −1
        for j = i to (m − 1) do
            sumDiv(i) := sumDiv(i) + y_j
        end for
        sumDiv(i) := (sumDiv(i) − 1)/(m − i)
        if sumDiv(i) ≥ y_(i−1) then
            th := sumDiv(i)
            exit for
        end if
    end for
    for i = 1 to m do
        b'_it := b'_it − th
        if y_it < 0 then
            b'_it := 0
        end if
    end for
    return b_t = b'_t
end function
```

allocation $\mathfrak{b}^{\omega}$, then the period wealth is denoted as $W_t(\mathfrak{b}^{\omega})$ to specify the exact portfolio allocation of the ω expert, i.e., CR-expert. An ALG which is based on information provided by the performance of the Ω experts on the market $\mathbf{x}$ is called experts ALG. Examples for experts algorithms are in [Cover, 1991] and [Gaivoronski and Stella, 2000]. Note that the practical calculation of an experts ALG can be a problem due to the exponential growth of the number of CR-experts in a market (see also Section 3.3). In this case, it is also possible to apply random CR-experts as described by Blum and Kalai [1999] and Ishijima [2001].

Window Size

Several algorithms take only time limited information into account. The idea is, younger information is more valuable than older one. Technically, a "sliding" window with given window size $w > 0$ is used. For example, the

ALG uses only information available from $t-w+1$ until t to generate $\mathbf{b}_{(t+1)}$. It should be noted, an ALG with window size always requires at the start of the investment horizon at least w trading periods before a calculation is possible. An example for an ALG with window size can be found in [Borodin et al., 2004], which considers even $2w$ past trading periods.

First and Second Order Information

The holding period return is defined as

$$x_t = \sum_{i=1}^{m} x_{it} b_{it} = \frac{W_t}{W_{(t-1)}} \tag{4.20}$$

and is showing the increase/decrease of wealth for one single trading period. Some algorithms exploit first and second order information, which are directly related to the current holding period return. The first order information is extracted by

$$\Theta_t^i = \frac{\partial \ln x_t}{\partial b_{it}} = \frac{x_{it}}{x_t} \tag{4.21}$$

and describes the price change of A_i in relation to the holding period return of the current portfolio allocation. A value $\Theta_t^i > 1$ ($\Theta_t^i < 1$) indicates that A_i performs during trading period t better (worse) than the current portfolio allocation $\mathbf{b}_t$. The second order information is calculated by

$$\Theta_t^{ij} = \frac{\partial^2 \ln x_t}{\partial b_{it} \partial b_{jt}} = -\frac{x_{it} x_{jt}}{x_t^2} \tag{4.22}$$

and is the combined price change of A_i and A_j for $i, j = 1, ..., m$ in relation to the quadratic holding period return. A value $\Theta_t^{ij} < 1$ ($\Theta_t^{ij} > 1$) quantifies that an equally weighted portfolio with only A_i and A_j performs during trading period t better (worse) than the current portfolio $\mathbf{b}_t$. An ALG using first order information is provided by Helmbold et al. [1998] and an ALG using first plus second order information by Agarwal et al. [2006].

Average Market Return

Some algorithms focus the market for one trading period as a totality. For this, let

$$\bar{x}_t = \frac{1}{m} \sum_{i=1}^{m} x_{it} \tag{4.23}$$

be the average market return (factor) during trading period t. The value for $\bar{x}_t$ can be used as a benchmark value. It is equal to the holding period return of a portfolio allocation with $b_{it} = \frac{1}{m}$ for all $i = 1, ..., m$, i.e., naive diversification. The algorithms provided by Li et al. [2012] are using this information $\bar{x}_t$.

Side Information

In general, algorithms use information directly or indirectly from the market $\mathbf{x}$. But Cover and Ordentlich [1996] showed how to create an ALG which takes external information of $\mathbf{x}$ into account. The so called side information is basically a variable with finite value range for $t = 1, ..., T$. Such information can arise in numerous ways. For example, for trading often the three forecasting signals "Buy", "Hold" or "Sell" are generated for a stock A_i, usually also based on fundamental information of the underlying company [for further explanations see Stickel, 1995]. Such signals could be the side information, where the value range in this case can be 1,2,3. An ALG with side information tries to find a dependency between $\mathbf{x}$ and the side information. If the forecasting signals are having real forecasting ability on the concrete market instance $\mathbf{x}$, then an ALG with side information could benefit from them.

4.2 Follow-the-Winner Algorithms

In the following a selection of algorithms which track either the BCR or the best expert is provided. A performance guarantee for each ALG without making any statistical assumptions can be given. In other words, any provided ALG has a lower bound on the terminal wealth in comparison to the BCR when the worst-case occurs (see also Section 3.5.5 for the concept of universality and Section 5.2 for the performance guarantee).

4.2.1 Successive Constant Rebalanced Algorithm

Consider the following strategy, an investor is starting for $\mathbf{b}_1$ with a naive diversification because in $t = 1$ no past information exists. In the subsequent trading periods the investor is searching for the best allocation $\mathfrak{b}^*$ when using a CR strategy for the investment horizon $1, ..., t$. Formally, the portfolio allocation is defined by

$$\mathbf{b}_{(t+1)}^{SCR} = \mathbf{b}_t^* = \arg \max_{\mathfrak{b} \in \mathfrak{B}_m} W_t^{CR}(\mathfrak{b}) \tag{4.24}$$

with a continuous simplex $\mathfrak{B}_m$. The ALG is called *successive constant rebalanced algorithm* (SCR) and is provided by Gaivoronski and Stella [2000]. In fact, the SCR works in the same way with a given set of Ω CR-experts, i.e., with a discretized simplex. The SCR chooses always the best CR-expert up to trading period t. The SCR for a discretized simplex is given by Algorithm 2.

Algorithm 2 Successive Constant Rebalanced Algorithm

function SCR(CR-experts)
    $\omega Best := 1$
    $WBest := 0$
    for $\omega = 1$ to Ω **do**
        if $W_t(\mathfrak{b}^\omega) > WBest$ **then**
            $\omega Best := \omega$
            $WBest := W_t(\mathfrak{b}^\omega)$
        end if
    end for
    $\mathbf{b}_{(t+1)} := \mathfrak{b}^{\omega Best}$
    return $\mathbf{b}_{(t+1)}$
end function

4.2.2 Universal Portfolio Algorithm

A robust procedure to set the portfolio proportion $\mathbf{b}_{(t+1)}$ is provided by Cover [1991] and denoted as *universal portfolio algorithm* (UP). Consider again a scenario with $\omega = 1, ..., \Omega$ CR-experts. Each CR-expert uses a unique $\mathfrak{b}^\omega$. To determine the proportion of wealth for A_i the UP considers from each expert the current period wealth W_t^ω and the fraction of A_i in the portfolio $\mathfrak{b}^\omega$. The idea is, if an asset is more beneficial than the other assets, then experts who have higher commitment in this asset should generate, on average, a greater period wealth than experts with less commitment.

Formally, with a continuous simplex $\mathfrak{B}_m$ the UP is defined as

$$\mathbf{b}_{(t+1)}^{UP} = \frac{\int \mathfrak{b} W_t(\mathfrak{b}) d\mathfrak{b}}{\int W_t(\mathfrak{b}) d\mathfrak{b}} \tag{4.25}$$

where $\mathbf{b}_1^{UP}$ is the naive diversification. If the simplex is discrete, such that Ω experts exist, then

$$b_{i(t+1)}^{UP} = \frac{\sum_{\omega=1}^{\Omega} \mathfrak{b}_i^{\omega} W_t(\mathfrak{b}^{\omega})}{\sum_{\omega=1}^{\Omega} W_t(\mathfrak{b}^{\omega})} \tag{4.26}$$

gives the proportion of UP for A_i with $i = 1, ..., m$. Algorithm 3 provides the UP based on Equation (4.26) as pseudocode when the period wealth of the Ω CR-experts is given. Note, the ALG of Vovk and Watkins [1998] is a generalization of UP, which allows an additional weighting of the experts.

Algorithm 3 Universal Portfolio Algorithm

function UP(CR-experts)
    $sum := 0$
    for $\omega = 1$ to Ω **do**
        $sum := sum + W_t(\mathfrak{b}^{\omega})$
    end for
    for $i = 1$ to m **do**
        $wSum := 0$
        for $\omega = 1$ to Ω **do**
            $wSum := wSum + \mathfrak{b}_i^{\omega} \times W_t(\mathfrak{b}^{\omega})$
        end for
        $b_{i(t+1)} := wSum/sum$
    end for
    return $\mathbf{b}_{(t+1)}$
end function

4.2.3 Exponential Gradient Algorithm

Helmbold et al. [1998] provide the *exponential gradient algorithm* (EG) which requires only a constant storage time per asset, i.e., linear time complexity (see Section 3.1.3). In contrast to SCR and UP, which are technically based on the information of experts, the EG is based on first order information (see also Equation (4.21)). In addition, $\mathbf{b}_1$ can be an arbitrary allocation from the simplex $\mathfrak{B}_m$. The proportion $\mathbf{b}_{(t+1)}$ is based on the current $\mathbf{b}_t$ by calculating the first order information. The idea is, if the return factor of A_i in trading period t is greater than the current period wealth change, then the proportion of A_i for $t+1$ should be increased. A learning rate η intensifies this effect. Thus, the EG is defined as

$$b_{i(t+1)}^{EG} = \frac{b_{it} \exp\left(\eta \Theta_t^i\right)}{\sum_{i=1}^m b_{it} \exp\left(\eta \Theta_t^i\right)} \tag{4.27}$$

for $i = 1, ..., m$. Note, if $\eta = 0$, the initial proportion in $t = 1$ remains the same for all subsequent trading periods. Another interesting aspect is, if $\eta < 0$, then the EG changes to a FTL algorithm. Algorithm 4 provides the EG as pseudocode.

Algorithm 4 Exponential Gradient Algorithm

function EG(η)
    $sum := 0$
    for $i = 1$ to m **do**
        $isum(i)$:=exp$((\eta \times x_{it})/x_t)$
        $sum := sum + isum(i)$
    end for
    for $i = 1$ to m **do**
        $b_{i(t+1)} := isum(i)/sum$
    end for
    return $\mathbf{b}_{(t+1)}$
end function

4.2.4 Online Newton Step Algorithm

Compared to the EG, the *online Newton step algorithm* (ONS) of Agarwal et al. [2006] takes additionally the second order information into account (see also Equation (4.22)). The ONS uses as parameters β and δ and is here presented in a simplified form as described in Li and Hoi [2014].

For each trading period, the matrix

$$\mathbf{A}_t = \begin{pmatrix} 1 - \sum_{\tau=1}^t \Theta_\tau^{11} & \dots & 0 - \sum_{\tau=1}^t \Theta_\tau^{1m} \\ \vdots & \ddots & \vdots \\ 0 - \sum_{\tau=1}^t \Theta_\tau^{m1} & \dots & 1 - \sum_{\tau=1}^t \Theta_\tau^{mm} \end{pmatrix} \tag{4.28}$$

is calculated where one component is denoted as a_t^{ij}. Note that $a_t^{ij} = 1 - \sum_{\tau=1}^{t} \Theta_\tau^{ij}$ is only on the diagonal of the matrix ($i = j$) and $a_t^{ij} = 0 - \sum_{\tau=1}^{t} \Theta_\tau^{ij}$ on the non-diagonal of the matrix ($i \neq j$). Let $\mathbf{A}_t^{-1}$ be the inverse of $\mathbf{A}_t$ where one component of the inverse is denoted as $\bar{a}_t^{ij}$. Obviously, this calculation step employs second order information as described in Section 4.1.3.

Consider the price change of each asset in comparison to the wealth change, i.e., first order information (see Equation (4.21)). The vector $\mathbf{o}_t$ combines the first and second order information by

$$\mathbf{o}_t = \begin{pmatrix} \delta(1+\frac{1}{\beta})\sum_{j=1}^{m} \bar{a}_t^{1j} \sum_{\tau=1}^{t} \Theta_\tau^j \\ \vdots \\ \delta(1+\frac{1}{\beta})\sum_{j=1}^{m} \bar{a}_t^{mj} \sum_{\tau=1}^{t} \Theta_\tau^j \end{pmatrix} \tag{4.29}$$

where one component of $\mathbf{o}_t$ is denoted as o_{it} with $i = 1, ..., m$. The allocation for $t+1$ is defined as

$$\mathbf{b}_{(t+1)}^{ONS} = \arg \min_{\mathfrak{b} \in \mathfrak{B}_m} (\mathbf{o}_t - \mathfrak{b})^{\mathrm{T}} \mathbf{A}_t (\mathbf{o}_t - \mathfrak{b}) \tag{4.30}$$

or rewritten as

$$\mathbf{b}_{(t+1)}^{ONS} = \arg \min_{\mathfrak{b} \in \mathfrak{B}_m} \left(\sum_{i=1}^{m} a_t^{ii}(o_{it} - \mathfrak{b}_i) + 2 \sum_{i=1}^{m-1} \sum_{j=i+1}^{m} a_t^{ij}(o_{it} - \mathfrak{b}_i)(o_{jt} - \mathfrak{b}_j) \right) \tag{4.31}$$

with $\mathbf{o}_t = (o_{1t} \cdots o_{mt})^{\mathrm{T}}$. Note, the minimization of Equation (4.31) cannot be solved by the projection given by Algorithm 1. Instead of, it is recommended to apply the methods described by Bertsekas [1982].

For an online calculation of the ONS use the following calculation steps. Start with

$$a_t^{ij} = a_{t-1}^{ij} - \Theta_t^{ij} \tag{4.32}$$

for $t = 1, ..., T$ and $i, j = 1, ..., m$ with $a_0^{ij} = 1$ when $i = j$ and otherwise $a_0^{ij} \neq 0$. The first order information is extracted by

$$o'_{it} = \left(1 + \frac{1}{\beta}\right) \Theta_t^i + o'_{i(t-1)} \tag{4.33}$$

with $o'_{i0} = 0$ for $i = 1, ..., m$. Then, combine first and second order information by

$$o_{it} = \delta \sum_{j=1}^{m} \bar{a}_t^{ij} o'_{jt} \tag{4.34}$$

for $i = 1, ..., m$. Algorithm 5 provides the ONS as pseudocode.

Algorithm 5 Online Newton Step Algorithm

function ONS(β, δ)
    for $i = 1$ to m **do**
        for $j = 1$ to m **do**
            Calculate a_t^{ij}
        end for
    end for
    Calculate $\mathbf{A}_t^{-1}$
    for $i = 1$ to m **do**
        Calculate o_{it}
    end for
    $\mathfrak{b}$ = solution of Equation (4.31)
    return $\mathbf{b}_{(t+1)} = \mathfrak{b}$
end function

4.3 Follow-the-Loser Algorithms

In the following selected algorithms from the type FTL are provided. The idea is based on the believe in mean reversion such that assets or experts with low performance in the past are going to perform high in the future and vice versa (for evidence in practice see also [DeBondt and Thaler, 1985], [Jegadeesh and Titman, 1993], [Richards, 1997] and [Malin and Bornholt, 2013]). According to Li and Hoi [2014, p. S.15-19], for FTL algorithms it is difficult to obtain a lower bound on the performance in comparison to a benchmark algorithm.

4.3.1 Anti Correlation Algorithm

Borodin et al. [2004] propose the *anti correlation algorithm* (AC) for the PSP. The underlying assumption of the AC is that all stocks within the market perform similarly in terms of the long term exponential growth rate. If an asset in a past time window shows a distinctly different performance than other assets, then this indicates a counter movement of performance in the future.

More precisely, consider two consecutive time windows, i.e., $k = 1, 2$. The first one ($k = 1$) contains the logarithmic return factors of all assets from $t = (t - 2w + 1), ..., (t - w)$. The second one ($k = 2$) contains the periods $t = (t - w + 1), ..., t$. Calculate the exponential growth rate for $A_1, ..., A_m$ in the time windows $k = 1, 2$ at the beginning of trading period t by

$$\hat{\mu}_i(w,t,k) = \frac{1}{w} \sum_{\tau=t-(3-k)w+1}^{t-(2-k)w} \ln x_{i\tau} \tag{4.35}$$

with a given window size $w \geq 2$. In addition, calculate the cross correlation of the logarithmic return factors between the two time windows for all combinations of $A_1, ..., A_m$ by

$$\check{\rho}_{i,j}(w,t) = \frac{\sum_{\tau=t-2w+1}^{t-w} \left[\ln(x_{i(\tau-w)}) - \hat{\mu}_i(w,t,1)\right]\left[\ln(x_{j(\tau+w)}) - \hat{\mu}_j(w,t,2)\right]}{\sqrt{\sum_{\tau=t-2w+1}^{t-w} \left[\ln(x_{i(\tau-w)}) - \hat{\mu}_i(w,t,1)\right]^2 \sum_{\tau=t-2w+1}^{t-w} \left[\ln(x_{j(\tau+w)}) - \hat{\mu}_j(w,t,2)\right]^2}} \tag{4.36}$$

with the assumption $\frac{0}{0} := 0$. Wealth is transferred from A_i to A_j when $\hat{\mu}_i(w,t,2) > \hat{\mu}_j(w,t,2)$ and $\check{\rho}_{i,j}(w,t) > 0$ is true. Then, calculate the claim for transferring wealth from A_i to A_j by

$$Claim_{i,j}(w,t) = \check{\rho}_{i,j}(w,t) - \left[\min(0, \check{\rho}_{i,i}(w,t)) + \min(0, \check{\rho}_{j,j}(w,t))\right] \tag{4.37}$$

otherwise $claim_{i,j}(w,t) := 0$. Obviously, the claim depends on the cross correlation between A_i and A_j plus the negative cross correlation with itself, A_i and A_i (A_j and A_j). Finally, the proportion for the beginning of trading period $t + 1$ is calculated by

$$b_{i(t+1)}^{AC} = b_{it}\left(1 - \sum_{j=1}^{m} \frac{Claim_{i,j}(w,t)}{\sum_{j'=1}^{m} Claim_{i,j'}(w,t)}\right) + \sum_{j=1}^{m} b_{jt}\left(\frac{Claim_{i,j}(w,t)}{\sum_{j'=1}^{m} Claim_{i,j'}(w,t)}\right) \tag{4.38}$$

with the assumption that $\frac{0}{0} := 0$. Algorithm 6 provides the AC as pseudocode.

Algorithm 6 Anti Correlation Algorithm

function AC(w)
    if $t < 2w$ **then**
        for $i = 1$ to m **do**
            $b_{i(t+1)} = \frac{1}{m}$
        end for
    else
        for $i = 1$ to m **do**
            $Mu_{i,1} := \hat{\mu}_i(w, t, 1)$
            $Mu_{i,2} := \hat{\mu}_i(w, t, 2)$
            for $j = 1$ to m **do**
                $Rho_{i,j} := \check{\rho}_{i,j}(w, t)$
            end for
        end for
        for $i = 1$ to m **do**
            for $j = 1$ to m **do**
                $claim_{i,j} := Claim_{ij}(w, t)$
            end for
        end for
        for $i = 1$ to m **do**
            $b_{i(t+1)} := DistributeClaims(i)$ ▷ Equation (4.38)
        end for
    end if
    return $\mathbf{b}_{(t+1)}$
end function

4.3.2 Passive Aggressive Mean Reversion Algorithm

The main idea of the *passive aggressive mean reversion algorithm* (PAMR), provided by Li et al. [2012], is to "punish" assets which showed in the past trading period a larger return factor than the average of the market. In other words, PAMR exploits the mean reversion of asset prices for one single trading period. Because strong falling return factors are sometimes rooted in a total or almost total crash of an asset (e.g., $x_{it} = 0$) the PAMR can alternate between an "aggressive" and "passive" investment strategy which is indicated by a loss function. The concept of switching strategies based on the input for the PAMR is based on the online passive aggressive learning technique from machine learning [for further explanations see Crammer et al., 2006].

Let $\hat{\bar{x}} \in \mathbb{R}^+$ be a given mean reversion parameter. The PAMR starts with $b_{it} = \frac{1}{m}$ for all $i = 1, ..., m$. Then, the sum of the square deviation from the average market change is calculated by

$$v_t = \sum_{i=1}^{m} (x_{it} - \bar{x}_t)^2 \tag{4.39}$$

and quantifies the current variation within the market only at trading period t. Then,

$$loss_t = \min\left\{0, \hat{\bar{x}} - x_t\right\}. \tag{4.40}$$

is a loss function. Obviously, if the holding period return x_t is greater than the given mean reversion level $\hat{\bar{x}}$, then the loss function becomes negative. The proportion for A_i at trading period $t+1$ is

$$b_{i(t+1)} = b_{it} + \lambda_t(loss_t, v_t) \times (x_{it} - \bar{x}_t) \tag{4.41}$$

where λ_t is a multiplier function. Note that, Li et al. [2012] construct the PAMR such that always $\lambda_t(\cdot, \cdot) \leq 0$ holds. Hence, assets with a price change greater than the market average are "punished". Depending on the calculation for λ_t, PAMR can be distinguished into the three types

$$\lambda_t(loss_t, v_t) = \begin{cases} \frac{loss_t}{v_t} & \text{(PAMR1)} \\ \min\left\{C, \frac{loss_t}{v_t}\right\} & \text{(PAMR2)} \\ \frac{loss_t}{v_t + \frac{1}{2C}} & \text{(PAMR3)} \end{cases} \tag{4.42}$$

where C is a given aggressiveness parameter (often assumed $C = 1.0$). If $\mathbf{b}_{(t+1)} \notin \mathfrak{B}_m$, then use Algorithm 1 to project $\mathbf{b}_{(t+1)}$ onto the simplex $\mathfrak{B}_m$. Algorithm 7 provides the PAMR as pseudocode.

4.3.3 Confidence Weighted Mean Reversion Algorithm

Li et al. [2013] provide the *confidence weighted mean reversion algorithm* (CWMR) as a further development of PAMR which applies the confidence weighted learning technique described in [Dredze et al., 2008] and [Crammer et al., 2008]. The underlying idea of the CWMR is to find the allocation with the minimum expectation for the holding period return with respect to the current return factors and a given confidence level. A minimization is executed because if in the market mean reversion exists, then a current minimum holding period return should revert in the subsequent

Algorithm 7 Passive Aggressive Mean Reversion Algorithm

function PAMR($\hat{\bar{x}}, \{C\}$)
    Calculate $\bar{x}_t$
    Calculate v_t
    Calculate $loss_t$
    for $i = 1$ to m **do**
        $b_{i(t+1)} := b_{it} + \lambda_t(loss_t, v_t) \times (x_{it} - \bar{x}_t)$ ▷ with given $\lambda_t(\cdot, \cdot)$
    end for
    $\mathbf{b}_{(t+1)} := Projection(\mathbf{b}_{(t+1)})$
    return $\mathbf{b}_{(t+1)}$
end function

trading period.

As parameters the CWMR requires a confidence parameter ϕ and a mean reversion parameter $\hat{\bar{x}}$. The parameter ϕ is similar to the bounds of a confidence interval in the normal distribution for a given confidence level α, i.e., $\phi \approx 2$ for $\alpha = 0.95$ [compare with Kanji, 2006, p. 188].

The CWMR algorithm exists in two types. One is based on the variance (CWMRV) and the other one on the standard deviation (CWMRS). Both start with $b_{i1} = \frac{1}{m}$ for $i = 1, ..., m$. Let v_{it} be a variation measure showing the variance of A_i at the end of trading period t for CWMRV and the standard deviation for CWMRS. Set initially $v_{i0} = \frac{1}{m^2}$ for CWMRV and $v_{i0} = \frac{1}{m}$ for CWMRS with $i = 1, ..., m$. The market variation at trading period t is quantified by

$$V_t = \sum_{i=1}^{m} x_{it}^2 v_{i(t-1)}. \tag{4.43}$$

Let

$$\tilde{x}_t = \frac{\sum_{i=1}^{m} x_{it} v_{i(t-1)}}{\sum_{i=1}^{m} v_{i(t-1)}} \tag{4.44}$$

be the confidence weighted average return factor for trading period t. Based on V_t and $\tilde{x}_t$ three auxiliary variables ($\mathsf{a}_t, \mathsf{b}_t$ and c_t) must be calculated, i.e.,

$$\mathsf{a}_t = \frac{2\phi {V_t}^2 - 2\phi V_t \tilde{x}_t \sum_{i=1}^m x_{it} v_{i(t-1)}}{x_t^2} \tag{4.45a}$$

$$\mathsf{b}_t = \frac{V_t - \tilde{x}_t \sum_{i=1}^m x_{it} v_{i(t-1)}}{x_t^2} + 2\phi V_t (\ln \hat{\tilde{x}} - \ln x_t) \tag{4.45b}$$

$$\mathsf{c}_t = \ln \hat{\tilde{x}} - \ln x_t - \phi V_t \tag{4.45c}$$

for CWMRV and

$$\mathsf{a}_t = \left(\frac{V_t - \tilde{x}_t \sum_{i=1}^m x_{it} v_{i(t-1)}}{x_t^2} + \frac{V_t \phi^2}{2} \right)^2 - \frac{V_t^2 \phi^4}{4} \tag{4.46a}$$

$$\mathsf{b}_t = 2 \left(\ln \hat{\tilde{x}} - \ln x_t \right) \left(\frac{V_t - \tilde{x}_t \sum_{i=1}^m x_{it} v_{i(t-1)}}{x_t^2} + \frac{V_t \phi^2}{2} \right) \tag{4.46b}$$

$$\mathsf{c}_t = \left(\ln \hat{\tilde{x}} - \ln x_t \right)^2 - V_t \phi^2 \tag{4.46c}$$

for CWMRS. Then, solve the equation

$$\mathsf{a}_t \lambda_t^2 + \mathsf{b}_t \lambda_t + \mathsf{c}_t = 0 \tag{4.47}$$

for λ_t by

$$\lambda_t = \begin{cases} \max\{\frac{-\mathsf{b}_t \pm \sqrt{\mathsf{b}_t^2 - 4\mathsf{a}_t \mathsf{c}_t}}{2\mathsf{a}_t}, 0\} & \text{if } \mathsf{a}_t \neq 0 \text{ and } \mathsf{b}_t^2 - 4\mathsf{a}_t\mathsf{c}_t \geq 0 \\ \max\{-\frac{\mathsf{c}_t}{\mathsf{b}_t}, 0\}, & \text{if } \mathsf{a}_t = 0 \text{ and } \mathsf{b}_t \neq 0 \\ 0 & \text{otherwise} \end{cases} \tag{4.48}$$

with the corresponding $\mathsf{a}_t, \mathsf{b}_t$ and c_t of CWMRS or CWMRV. After this, calculate

$$v'_{it} = \frac{1}{\frac{1}{v_{i(t-1)}} + 2\lambda_t \phi x_{it}^2} \tag{4.49}$$

for CWMRV and

$$v'_{it} = \frac{1}{\frac{1}{v_{i(t-1)}} + \frac{\lambda_t}{\sqrt{U_t}} \phi x_{it}^2} \tag{4.50}$$

for CWMRS with

$$\sqrt{U_t} = \frac{-\lambda_t \phi V_t + \sqrt{(\lambda_t \phi V_t)^2 + 4V_t}}{2}. \tag{4.51}$$

The normalized deviation of A_i at the end of trading period t is then determined by

$$v_{it} = \frac{v'_{it}}{m^2 \sum_{i=1}^{m} v'_{it}}. \tag{4.52}$$

The proportion of A_i for trading period $t+1$ is calculated by

$$b_{i(t+1)} = b_{it} - \lambda_t v_{it} \frac{x_{it} - \tilde{x}_t}{x_t} \tag{4.53}$$

for all $i = 1, ..., m$. If $\mathbf{b}_{(t+1)} \notin \mathfrak{B}_m$, then use Algorithm 1 to project $\mathbf{b}_{(t+1)}$ onto the simplex $\mathfrak{B}_m$. Algorithm 8 provides the CWMR as pseudocode.

Algorithm 8 Confidence Weighted Mean Reversion Algorithm

function CWMR($\phi, \hat{\tilde{x}}$)
    for $i = 1$ to m **do**
        Calculate $v_{i(t-1)}$
    end for
    Calculate x_t
    Calculate V_t
    Calculate $\tilde{x}_t$
    Calculate a_t, b_t and c_t ▷ for CWMRV or CRMRS
    Calculate λ_t
    for $i = 1$ to m **do**
        $b_{i(t+1)} := b_{it} - \lambda_t v_{i(t-1)} \times ((x_{it} - \tilde{x}_t)/x_t)$
    end for
    $\mathbf{b}_{(t+1)} := Projection(\mathbf{b}_{(t+1)})$
    return $\mathbf{b}_{(t+1)}$
end function

4.3.4 Online Moving Average Mean Reversion Algorithm

The previous described online algorithms PAMR and CWMR exploit mean reversion based on the return factors for only one past trading period. In contrast, the *online moving average mean reversion algorithm* (OLMAR) of Li and Hoi [2012] is an extension of PAMR and it considers more than one past trading periods. The average of asset prices in a past time window is employed as a predictor for the asset price in the next trading period.

Consider the initial price of A_i at the beginning of trading period t is $q_{i0} = 1$ for all $i = 1, ..., m$. The price of any asset at the end of trading period t is determined by $q_{it} = q_{i(t-1)}x_{it}$. Let w be a given window size and $\hat{\bar{x}}$ a given parameter for mean reversion. Then, the moving average is calculated for $\mathbf{b}_{(t+1)}$ with $t \geq w$ by

$$MA_{it}^{w} = \frac{\sum_{\tau=t-w+1}^{t} q_{i\tau}}{w} \tag{4.54}$$

and is combined with the current price q_{it} by

$$\bar{x}_{it}^{w} = \frac{MA_{it}^{w}}{q_{it}} = \frac{\sum_{\tau=t-w+1}^{t} q_{i\tau}}{wq_{it}} \tag{4.55}$$

at the end of trading period t. The variable $\bar{x}_{it}^{w}$ quantifies whether the current price of A_i is greater ($\bar{x}_{it}^{w} < 1$) or lower ($\bar{x}_{it}^{w} > 1$) than the current moving average. Consider

$$\bar{x}_{t}^{w} = \frac{1}{m}\sum_{i=1}^{m} \bar{x}_{it}^{w} \tag{4.56}$$

as the market average of all $\bar{x}_{it}^{w}$. In a next step determine λ_t

$$\lambda_t = \max\left\{ 0, \frac{\hat{\bar{x}} - \sum_{i=1}^{m} b_{it}\bar{x}_{it}^{w}}{\sum_{i=1}^{m}(\bar{x}_{it}^{w} - \bar{x}_{t}^{w})^2} \right\} \tag{4.57}$$

with $\frac{\hat{\bar{x}} - \sum_{i=1}^{m} b_{it}\bar{q}_{it}^{w}}{0} = 0$.
The portfolio allocation for trading period $t+1$ is calculated by

$$b_{i(t+1)} = b_{it} + \lambda_t(\bar{x}_{it}^{w} - \bar{x}_{t}^{w}) \tag{4.58}$$

and requires a projection onto the simplex $\mathfrak{B}_m$ by Algorithm 1 if $\mathbf{b}_{(t+1)} \notin \mathfrak{B}_m$. Algorithm 9 provides the OLMAR as pseudocode.

4.3.5 Robust Median Reversion Algorithm

The *robust median reversion algorithm* (RMR) is a further development of the OLMAR and is provided by Huang et al. [2013]. The OLMAR also exploits mean reversion with a multi-period perspective. In general, RMR is similar to OLMAR. But, instead of using the average, the RMR uses a vector based median of asset prices as a predictor for the next trading period.

Algorithm 9 Online Moving Average Mean Reversion Algorithm

function OLMAR($w, \hat{\bar{x}}$)
    for $i = 1$ to m **do**
        Calculate $\bar{x}_{it}^{w}$
    end for
    Calculate $\bar{x}_{t}^{w}$
    Calculate λ_t
    for $i = 1$ to m **do**
        $b_{i(t+1)} := b_{it} + \lambda_t \times (\bar{x}_{it}^{w} - \bar{x}_{t}^{w})$
    end for
    $\mathbf{b}_{(t+1)} := Projection(\mathbf{b}_{(t+1)})$
    return $\mathbf{b}_{(t+1)}$
end function

As input parameters a window size w, a mean reversion parameter $\hat{\bar{x}}$, a threshold value $\hat{\tau}$ and an iteration parameter K must be given.

Let $\hat{q}_{it}^{w}$ be an estimator for the next asset price of A_i (i.e., $q_{i(t+1)}$) based on information within a window size w. Let

$$\hat{\mathbf{q}}_{t}^{w} = (\hat{q}_{1t}^{w} \cdots \hat{q}_{mt}^{w}) \tag{4.59}$$

be a vector consisting of all price estimators. The values of the vector are determined by iteration. Consider $k = 1, ..., K$ iterations where the iteration k of A_i is denoted as $\hat{q}_{it}^{w}(k)$ for $i = 1, ..., m$. The first iteration $k = 1$ is set to

$$\hat{q}_{it}^{w}(1) = MA_{it}^{w} \tag{4.60}$$

for $i = 1, ..., m$ (see Equation (4.54)). Then, calculate the Euclidean distance for $\tau = 0, ..., (w-1)$ by

$$\delta_{t\tau}^{w}(k) = \|\mathbf{q}_{(t-\tau)} - \hat{\mathbf{q}}_{t}^{w}(k)\| = \sqrt{\sum_{i=1}^{m} \left(q_{i(t-\tau)} - \hat{q}_{it}^{w}(k)\right)^2} \tag{4.61}$$

with $\mathbf{q}_{(t-\tau)}$ as the price vector of the market at the end of trading period $(t-\tau)$. The $\delta_{t\tau}^{w}(k)$ is in the following calculations only used in the denominator.

The Euclidean distance is modified by

$$\delta_{t\tau}^{\prime w}(k) = \begin{cases} \frac{1}{\delta_{t\tau}^{w}(k)} & \text{if } \hat{\mathbf{q}}_t^w(k) \neq \mathbf{q}_{(t-\tau)} \\ 0 & \text{if } \hat{\mathbf{q}}_t^w(k) = \mathbf{q}_{(t-\tau)} \end{cases} \tag{4.62}$$

to prevent a division by zero in the following. Then, calculate

$$\kappa_t^w(k) = \begin{cases} 1 & \text{if } \hat{\mathbf{q}}_t^w(k) = \mathbf{q}_{(t-\tau)}, \tau = 0, ..., w-1 \\ 0 & \text{if } \hat{\mathbf{q}}_t^w(k) \neq \mathbf{q}_{(t-\tau)} \end{cases} \tag{4.63}$$

and

$$\gamma_t^w(k) = \sqrt{\sum_{i=1}^{m} \left(\sum_{\tau=0}^{w-1} \left(q_{i(t-\tau)} - \hat{q}_{it}^w \right) \delta_{t\tau}^{\prime w}(k) \right)^2} \tag{4.64}$$

and

$$\zeta_{it}^w(k) = \frac{\sum_{l=0}^{w-1} q_{i(t-l)} \delta_{tl}^{\prime w}(k)}{\sum_{l=0}^{w-1} \delta_{tl}^{\prime w}(k)} \tag{4.65}$$

for $i = 1, ..., m$. The $\kappa_t^w(k)$, $\gamma_t^w(k)$ and the $\zeta_{1t}^w(k), ..., \zeta_{mt}^w(k)$ are used to determine the next iteration $k+1$ by

$$\hat{q}_{it}^w(k+1) = \begin{cases} \max\left(0, 1 - \frac{\kappa_t^w(k)}{\gamma_t^w(k)}\right) \zeta_{it}^w(k) + \min\left(1, \frac{\kappa_t^w(k)}{\gamma_t^w(k)}\right) \hat{q}_{it}^w(k) & \text{if } \gamma_t^w(k) \neq 0 \\ \hat{q}_{it}^w(k) & \text{otherwise} \end{cases} \tag{4.66}$$

for $i = 1, ..., m$. The iteration stops when either $k = K$ or a toleration level τ_k is passed, i.e., $\tau_k \leq \hat{\tau}$ with given threshold value $\hat{\tau}$ for τ_k. The following equation quantifies

$$\tau_k = \frac{\|\hat{q}_{it}^w(k+1) - \hat{q}_{it}^w(k)\|}{\|\hat{q}_{it}^w(k+1)\|} = \sqrt{\frac{\sum_{i=1}^{m} (\hat{q}_{it}^w(k) - \hat{q}_{it}^w(k+1))^2}{\sum_{i=1}^{m} (\hat{q}_{it}^w(k))^2}} \tag{4.67}$$

this toleration level. Finally, set $\hat{\mathbf{q}}_t^w = \hat{\mathbf{q}}_t^w(k+1)$.

The following steps are analogous to the steps of the OLMAR. Determine

$$\hat{x}_{it}^w = \frac{\hat{q}_{it}^w}{q_{it}} \tag{4.68}$$

as an indicator whether the predicted price $\hat{q}_{it}^w$ will increase or decrease in comparison to the current price q_{it}. Then, determine

$$\hat{x}_t^w = \frac{1}{m}\sum_{i=1}^{m}\hat{x}_{it}^w \tag{4.69}$$

and in a next step calculate λ_t by

$$\lambda_t = \max\left\{0, \frac{\hat{\bar{x}} - \sum_{i=1}^{m} b_{it}\hat{x}_{it}^w}{\sum_{i=1}^{m}(\hat{x}_{it}^w - \hat{x}_t^w)^2}\right\} \tag{4.70}$$

with $\frac{\hat{\bar{x}} - \sum_{i=1}^{m} b_{it}\hat{x}_{it}^w}{0} := 0$. The portfolio allocation for trading period $t+1$ is calculated by

$$b_{it+1} = b_{it} + \lambda_t(\hat{x}_{it}^w - \hat{x}_t^w) \tag{4.71}$$

and requires a projection onto the simplex $\mathfrak{B}_m$ by Algorithm 1 if $\mathbf{b}_{(t+1)} \notin \mathfrak{B}_m$. Algorithm 10 provides the RMR as pseudocode.

4.4 Conclusions

Summing up, a survey of online algorithms from the type of FTW and FTL with a technical perspective is given in this chapter. The understanding of the construction of the considered online algorithms should be improved. Hence, the implementation of these algorithms in a software tool is facilitated. The distinction between FTW and FTL is clarified by the support of a virtual market.

However, a satisfying performance evaluation of the considered online algorithms is missing. A first indicator is the time complexity of each ALG which is shown in Table 4.3 [taken from Huang et al., 2013, p. 2010]. Algorithms from type FTL have a lower time complexity than FTW and most of them require linear computing time. In contrast, there are FTW algorithms with an exponential time complexity. In practice, this can be a problem (see Section 3.1.3).

In addition, the optimization function for one single trading period (in the literature denoted as capital growth form) for each of the selected algorithms is provided in its essential idea and taken from [Li and Hoi, 2014, pp. 23-24]. It shows the impact of the change of portfolio allocation $\mathbf{b}_t$ between two consecutive trading periods. The function $R(\mathfrak{b}, \mathbf{b}_t)$ and $R(\mathfrak{b})$ represent regularization terms. They control the distance between the current and next portfolio allocation. Based on the temporal origin of the required

Algorithm 10 Robust Median Reversion Algorithm

```
function RMR($w, \hat{\hat{x}}, \hat{\tau}, K$)
    for $i = 1$ to $m$ do
        Calculate $\hat{q}_{it}^{w}(1)$
    end for
    for $k = 2$ to $K$ do
        for $l = 0$ to $w - 1$ do
            Calculate $\delta_{tl}^{w}(k)$
            Calculate $\delta_{tl}^{'w}(k)$
        end for
        Calculate $\gamma_t^{w}(k)$
        Calculate $\kappa_t^{w}(k)$
        for $i = 1$ to $m$ do
            Calculate $\zeta_{it}^{w}(k)$
            Calculate $\hat{q}_{it}^{w}(k)$
        end for
        Calculate $\tau_k$
        if $\tau_k \leq \hat{\tau}$ then
            exit for
        end if
    end for
    Calculate $\hat{x}_t^{w}$
    Calculate $\lambda_t$
    for $i = 1$ to $m$ do
        $b_{i(t+1)} := b_{it} + \lambda_t \times (\hat{x}_{it}^{w} - \hat{x}_t^{w})$
    end for
    $\mathbf{b}_{(t+1)} := Projection(\mathbf{b}_{(t+1)})$
    return $\mathbf{b}_{(t+1)}$
end function
```

ALG	Time complexity	Capital growth form
SCR	$O(T^m)$	$\mathbf{b}_{t+1} = \arg\max_{\mathfrak{b}\in\mathfrak{B}_m} \frac{1}{t}\sum_{\tau=1}^{t} \ln \sum_{i=1}^{m} \mathfrak{b}_i x_{i\tau}$
UP	$O(T^m)$	$\mathbf{b}_{t+1} = \arg\max_{\mathfrak{b}\in\mathfrak{B}_m} f(\mathfrak{b}, \mathbf{x}_1, ..., \mathbf{x}_t)$
EG	$O(Tm)$	$\mathbf{b}_{t+1} = \arg\max_{\mathfrak{b}\in\mathfrak{B}_m} \ln \sum_{i=1}^{m} \mathfrak{b}_i x_{it} - \lambda R(\mathfrak{b}, \mathbf{b}_t)$
ONS	$O(Tm^3)$	$\mathbf{b}_{t+1} = \arg\max_{\mathfrak{b}\in\mathfrak{B}_m} \frac{1}{t}\sum_{\tau=1}^{t} \ln \sum_{i=1}^{m} \mathfrak{b}_i x_{i\tau} - \lambda R(\mathfrak{b})$
AC	$O(Tm^2)$	$\mathbf{b}_{t+1} = \arg\min_{\mathfrak{b}\in\mathfrak{B}_m} f(\mathfrak{b}, \mathbf{x}_{(t-2w+1)}, ..., \mathbf{x}_t)$
PAMR	$O(Tm)$	$\mathbf{b}_{t+1} = \arg\min_{\mathfrak{b}\in\mathfrak{B}_m} \sum_{i=1}^{m} \mathfrak{b}_i x_{it} + \lambda R(\mathfrak{b}, \mathbf{b}_t)$
CWMR	$O(Tm)$	$\mathbf{b}_{t+1} = \arg\min_{\mathfrak{b}\in\mathfrak{B}_m} f(\sum_{i=1}^{m} \mathfrak{b}_i x_{it}) + \lambda R(\mathfrak{b}, \mathbf{b}_t)$
OLMAR	$O(Tm)$	$\mathbf{b}_{t+1} = \arg\max_{\mathfrak{b}\in\mathfrak{B}_m} \sum_{i=1}^{m} \mathfrak{b}_i \bar{x}_{it}^{w} - \lambda R(\mathfrak{b}, \mathbf{b}_t)$
RMR	$O(Tm)+O(TK)$	$\mathbf{b}_{t+1} = \arg\max_{\mathfrak{b}\in\mathfrak{B}_m} \sum_{i=1}^{m} \mathfrak{b}_i \hat{x}_{it}^{w} - \lambda R(\mathfrak{b}, \mathbf{b}_t)$

Table 4.3: Summary of time complexity [Huang et al., 2013, p. 2010] and capital growth form [Li and Hoi, 2014, p. 23] for selected algorithms (UP and AC are added)

information, the capital growth form of the selected algorithms can be distinguished into three categories:

(i) **Current period**: The next portfolio allocation depends only on information occurred in the current trading period, e.g., EG, PAMR, CWMR.

(ii) **Last w periods**: The next portfolio allocation depends only on information occurred in the last w (or $2w$) trading periods, e.g., AC, OLMAR, RMR.

(iii) **All periods**: The next portfolio allocation depends on information taken from all available trading periods from the past, i.e., $\tau = 1, ..., t$. For example, see SCR, UP and ONS.

Several investigations indicate that online algorithms from the type FTL are superior to algorithms from the type FTW concerning the maximization of terminal wealth. For examples, consider the results of the selected online algorithms on the four sample markets given in Table 4.4 which are often used in the literature, e.g., [Cover, 1991], [Helmbold et al., 1998] and [Agarwal et al., 2006]. The results extremely differ and are encouraging. But, there are still two open issues. In line with [Borodin and El-Yaniv, 1998, pp. 309-311] and [Li and Hoi, 2014, p. 29], an online ALG that incorporates risk management and act in the same manner as the considered FTW or FTL algorithms is missing. In addition, the empirical performance of the considered online algorithms is not satisfactorily discussed (see Chapter 6.1.2).

ALG	coke & ibm	comme & kinar	comme & meico	iro & kinar
SCR	5.56	28.25	31.27	16.81
UP	14.18	78.47	72.63	38.67
EG	14.90	110.96	94.28	64.43
ONS	18.12	442.05	86.28	32.75
AC	158.22	1.05E+06	6.96E+02	1.29E+06
PAMR	18.68	1.56E+11	7.16E+03	1.68E+11
CWMR	19.46	1.14E+11	5.39E+03	1.17E+11
OLMAR	35.70	8.21E+13	2.54E+05	2.05E+12
RMR	37.42	7.27E+13	8.19E+04	4.02E+12

Table 4.4: Summary of W_T with $W_0 = 1$ for selected online algorithms with daily rebalancing achieved on four sample markets with $m = 2$ and $T = 5651$; The parameters of each ALG are given in Section 6.1.1 and the calculation is done with the software tool which is introduced in Chapter 7.

5 Proposed Algorithms with Risk Management

In this chapter two new online algorithms for the portfolio selection problem are proposed. They contribute to close the gap between the finance community and the machine learning community. The algorithms incorporate risk management, but also have a provable lower bound on the terminal wealth in a worst-case scenario. The results[4] of this chapter are already presented, submitted and published at international conferences[5] and conference proceedings[6].

5.1 Preliminaries

Given is a risky market $\mathbf{x} = \mathbf{x}_1, ..., \mathbf{x}_T$ with $i = 1, ..., m$ assets and $t = 1, ..., T$ trading periods. Each trading period consists of a vector with $i = 1, ..., m$ return factors, i.e., $\mathbf{x}_t = (x_{1t}, ..., x_{mt})$. To invest into the market $\mathbf{x}$ with an initial wealth $W_0 = 1$ an ALG produces the corresponding sequence of portfolio allocations $\mathbf{b} = \mathbf{b}_1, ..., \mathbf{b}_T$ with $\mathbf{b}_t = (b_{1t}, ..., b_{mt})$.

In the MLC, when focusing on the PSP, the objective is to maximize terminal wealth without making any statistical assumptions on the market [compare with Li and Hoi, 2014, p. 2], cf. Mathematical Program 9. The most noted online ALG solving the PSP is the UP, which is introduced by Cover [1991]. According to [Borodin and El-Yaniv, 1998, pp. 309-311] and [Li and Hoi, 2014, p. 29], online algorithms with risk management are missing.

---

[4]RAPS and CRAPS are developed and empirically tested by Robert Dochow. The competitive analysis is done in joint work with Esther Mohr and Günter Schmidt.

[5]RAPS is presented at the OR2013 by Robert Dochow; RAPS and CRAPS are presented at the APMOD2014 by Esther Mohr.

[6]RAPS is published at [Dochow et al., 2014]; RAPS and CRAPS are submitted to a special issue of Annals of Operations Research in joint work with Esther Mohr.

Mathematical Program 9 - offline PSP

Given: $\mathbf{x}$
Find: $\mathbf{b}$

$$
\begin{array}{lll}
\textbf{max} & W_T = W_0 \prod_{t=1}^{T} \sum_{i=1}^{m} x_{it} b_{it} & \\
\textbf{s. t.} & \text{(I)} \quad \sum_{i=1}^{m} b_{it} = 1 & , t = 1, ..., T \\
 & \text{(II)} \quad b_{it} \geq 0 & , t = 1, ..., T, \; i = 1, ..., m
\end{array}
$$

In contrast, the FC formulates statistical assumptions for the market and takes risk management into account when making a decision for the PSP (see [Modigliani and Modigliani, 1997] and [Borodin and El-Yaniv, 1998, p. 309]). On the one side it is done directly by using a measure of risk which is minimized subject to several constraints for each trading period [e.g., the decision model of Markowitz, 1952]. On the other side a measure of risk is used in an indirect way, by maximizing a measure for the risk-adjusted performance, which in turn also consists of a risk measure [e.g., the decision model of Sharpe, 1966].

Technically, the employment of risk management can be explained as follows. In general, the objective of an investor is to maximize the terminal wealth W_T as described in the previous mathematical program. But, the market $\mathbf{x}$ is risky and therefore only known in hindsight. To avoid the loss of wealth the investor wants to keep under control the variation of $W_1, ..., W_T$. The investor is choosing $\mathbf{b}$ such that a W_T as high as possible with respect to the variation of $W_1, ..., W_T$ is realized. This can be done by taking a direct or indirect measure of risk into account, for each decision during the investment horizon of the PSP. Examples of the FC, where one $\mathbf{b}_t$ is derived, with incorporating risk management, are with respect to different statistical assumptions to find in [Markowitz, 1952], [Tobin, 1958], [Sharpe, 1963], [Mossin, 1966], [Lee, 1977], [Kraus and Litzenberger, 1976], [Garlappi et al., 2007] and [Hoe et al., 2010].

In this chapter two new online algorithms are proposed which incorporate risk management. These do not require any statistical assumptions for the market and keep under control the variation of the period wealth $W_1, ..., W_T$. The algorithms are denoted as *risk-adjusted portfolio selection algorithm* (RAPS) and *combined risk-adjusted portfolio selection algorithm* (CRAPS). RAPS and CRAPS are both experts algorithms and are motivated by the prominent UP. Although RAPS and CRAPS incorporate a measure of risk, but still an analytically performance guarantee for W_T can be given. The bounds for W_T are only valid in a worst-case scenario with $x_{it} \leq 1$ for all $i = 1, ..., m$ and $t = 1, ..., T$.

The underlying idea of RAPS and CRAPS is as follows. Consider $\omega = 1, ..., \Omega$ experts, each uses an own unique ALG on the market $\mathbf{x}$ (all starts with $W_0 = 1$). Consider the historical period wealth of all experts and determine for each the minimum period wealth m_t^ω and maximum period wealth M_t^ω up to time t. Then, calculate the maximum historical fluctuation by $\frac{m_t^\omega}{M_t^\omega}$ as an indicator for the risk of the expert ω. In fact, the $\frac{m_t^\omega}{M_t^\omega}$ is a simplified measure for the maximum drawdown (compare with Section 3.2.2). If MDD^ω is the maximum drawdown of expert ω, then in a Kelly market (see Section 3.5.5) it holds $\frac{m_T^\omega}{M_T^\omega} = 1 - MDD^\omega$. Nevertheless, experts with large (low) $\frac{m_t^\omega}{M_t^\omega}$ in trading period t indicate low (large) risk. RAPS and CRAPS overweight assets used by experts with low risk. One may ask about the relation between the risk of experts during the investment horizon and the risk of RAPS and CRAPS at the end of the investment horizon. It is assumed that experts with low (high) risk in the past, have low (high) risk in the future. An ALG whose rebalancing over the investment horizon is based on experts with low (high) risk will possess a low (high) risk at the end of the investment horizon.

Due to the fact that in a Kelly market $W_t^\omega = \frac{m_t^\omega}{M_t^\omega}$ because $M_t^\omega = W_0 = 1$ it is possible to give a worst-case performance guarantee for RAPS and CRAPS with respect to W_T. The lower bounds of W_T are only valid when a Kelly market occurs. In this market, the allocation of RAPS is always equal to UP. In addition, CRAPS put always less proportion in the more losing assets such that the loss of CRAPS will never be smaller than that of UP and RAPS. Although it would be sufficient to show these findings, but for the sake of completeness the comprehensive proofs for the bounds of RAPS and CRAPS as in the literature for UP are performed.

5.1.1 Worst-Case Logarithmic Wealth Ratio

In line with Cesa-Bianchi and Lugosi [2006, pp. 276-282] and Section 3.5, assume the performance of an online ALG applied by an investor is measured in terminal wealth W_T and compared with the performance of a benchmark ALG executed by an expert in a worst-case scenario, i.e., execution of a competitive analysis. For this, consider B^ω as one benchmark ALG executed by the expert ω from a finite set of possible benchmark algorithms $\mathfrak{B}$, i.e., $B^\omega \in \mathfrak{B}$ with $\omega = 1, ..., \Omega$ benchmark algorithms and $\omega = 1, ..., \Omega$ experts. In fact, the ALG competes against all benchmark algorithms in $\mathfrak{B}$.

Let $W_T(ALG, \mathbf{x})$ be the terminal wealth generated through the application of ALG on the market instance $\mathbf{x}$. The corresponding terminal wealth of the expert ω that applies B^ω on the same market instance is denoted as $W_T(B^\omega, \mathbf{x})$. If the benchmark ALG is a constant rebalancing ALG with fix portfolio allocation $\mathfrak{b}^\omega$, then the terminal wealth of the specific benchmark ALG is denoted as $W_T(\mathfrak{b}^\omega, \mathbf{x})$ (CR-expert).

The concepts to evaluate the performance of an online ALG can be distinguished into four cases:

(i) The performance of an online ALG is expressed by the worst-case logarithmic wealth ratio with a finite number of benchmark algorithms

$$\mathbf{W}_T(ALG, \mathfrak{B}) = \max_{\mathbf{x}} \max_{B^\omega \in \mathfrak{B}} \ln \frac{W_T(B^\omega, \mathbf{x})}{W_T(ALG, \mathbf{x})} \tag{5.1}$$

for arbitrary instances of $\mathbf{x}$. In other words, it is searched for the maximum ratio between $W_T(B^\omega, \mathbf{x})$ and $W_T(ALG, \mathbf{x})$ when B^ω and $\mathbf{x}$ can be chosen arbitrarily.

(ii) If $\mathbf{x}$ is limited to a specific class of market instances $\mathbf{X}$, then the worst-case logarithmic wealth ratio with a finite number of benchmark algorithms and limited markets is defined as

$$\mathbf{W}_T(ALG, \mathfrak{B}, \mathbf{X}) = \max_{\mathbf{x} \in \mathbf{X}} \max_{B^\omega \in \mathfrak{B}} \ln \frac{W_T(B^\omega, \mathbf{x})}{W_T(ALG, \mathbf{x})}. \tag{5.2}$$

The concepts (i) and (ii) are extended by considering only constant rebalancing algorithms as benchmarks executed by CR-experts. Each CR-expert is represented by one point on the simplex

$$\mathfrak{B}_m = \left(\mathfrak{b} \in \mathfrak{R}^m : \mathfrak{b}_i \geq 0, \sum_{i=1}^{m} \mathfrak{b}_i = 1 \right) \tag{5.3}$$

as discussed in Section 3.3. Since the simplex $\mathfrak{B}_m$ is continuous an infinite number of CR-experts can be generated.

(iii) From this perspective, the worst-case logarithmic wealth ratio with an infinite number of benchmark algorithms is defined as

$$\mathbf{W}_T(ALG, \mathfrak{B}_m) = \max_{\mathbf{x}} \max_{\mathfrak{b} \in \mathfrak{B}_m} \ln \frac{W_T(\mathfrak{b}, \mathbf{x})}{W_T(ALG, \mathbf{x})} \tag{5.4}$$

(iv) and the worst-case logarithmic wealth ratio with an infinite number of benchmark algorithms and limited markets as

$$\mathbf{W}_T(ALG, \mathfrak{B}_m, \mathbf{X}) = \max_{\mathbf{x}\in\mathbf{X}} \max_{\mathfrak{b}\in\mathfrak{B}_m} \ln \frac{W_T(\mathfrak{b}, \mathbf{x})}{W_T(ALG, \mathbf{x})}. \tag{5.5}$$

Note that the worst-case logarithmic wealth ratio is a synonym for the *regret* as described in Section 3.5.5. If the logarithmic calculation is excluded, then the problem is reduced to find a competitive ratio c as described in the Sections 3.5.1, 3.5.2 and 3.5.3.

5.1.2 Universal Portfolio Algorithm

The UP of Cover [1991] is defined as

$$b_{i(t+1)}(UP) = \frac{\sum_{\omega=1}^{\Omega} \mathfrak{b}_i^{\omega} W_t(\mathfrak{b}^{\omega}, \mathbf{x})}{\sum_{\omega=1}^{\Omega} W_t(\mathfrak{b}^{\omega}, \mathbf{x})} \tag{5.6}$$

with a finite number of CR-experts and is solving the problem described by Mathematical Program 9. In contrast, for an infinite number of CR-experts the UP is defined as

$$\mathbf{b}_{(t+1)}(UP) = \frac{\int \mathfrak{b} W_t(\mathfrak{b}, \mathbf{x}) d\mathfrak{b}}{\int W_t(\mathfrak{b}, \mathbf{x}) d\mathfrak{b}}. \tag{5.7}$$

Thus, UP considers in each trading period the period wealth of each expert to maximize the terminal wealth of the investor. One property of the UP is the universality. An ALG is universal, if its exponential growth rate converges asymptotically to the exponential growth rate of the best expert for increasing T, i.e.,

$$\frac{1}{T} \ln W_T(ALG\mathbf{x}) - \frac{1}{T} \ln max_{B^{\omega}\in\mathfrak{B}} W_T(B^{\omega}, \mathbf{x}) \longrightarrow 0 \tag{5.8}$$

or

$$\frac{1}{T} \ln W_T(ALG, \mathbf{x}) - \frac{1}{T} \ln max_{\mathfrak{b}\in\mathfrak{B}_m} W_T(\mathfrak{b}, \mathbf{x}) \longrightarrow 0 \tag{5.9}$$

as described in Section 3.5.5.

In fact, the terminal wealth of UP is always the average of all experts.

Consider Ω CR-experts, it is easy to see that

$$\begin{aligned}
W_T(UP, \mathbf{x}) &= \prod_{t=1}^{T}\sum_{i=1}^{m} x_{it} b_{it}(UP) \\
&= \prod_{t=1}^{T}\sum_{i=1}^{m} \frac{\sum_{\omega=1}^{\Omega} \mathfrak{b}_i^{\omega} W_{(t-1)}(\mathfrak{b}^{\omega}, \mathbf{x}) x_{it}}{\sum_{\omega=1}^{\Omega} W_{(t-1)}(\mathfrak{b}^{\omega}, \mathbf{x})} \\
&= \prod_{t=1}^{T} \frac{\sum_{\omega=1}^{\Omega}\sum_{i=1}^{m} \mathfrak{b}_i^{\omega} W_t(\mathfrak{b}^{\omega}, \mathbf{x})}{\sum_{\omega=1}^{\Omega} W_{(t-1)}(\mathfrak{b}^{\omega}, \mathbf{x})} \\
&= \prod_{t=1}^{T} \frac{\sum_{\omega=1}^{\Omega} W_t(\mathfrak{b}^{\omega}, \mathbf{x})}{\sum_{\omega=1}^{\Omega} W_{(t-1)}(\mathfrak{b}^{\omega}, \mathbf{x})} \\
&= \frac{\sum_{\omega=1}^{\Omega} W_T(\mathfrak{b}^{\omega}, \mathbf{x})}{\sum_{\omega=1}^{\Omega} W_0} \\
&= \frac{1}{\Omega}\sum_{\omega=1}^{\Omega} W_T(\mathfrak{b}^{\omega}, \mathbf{x})
\end{aligned} \tag{5.10}$$

with $W_0 = 1$. Thus, UP is performing a buy-and-hold of all experts [compare with Cesa-Bianchi and Lugosi, 2006, p. 283].

As described in Section 3.5.5, any universal ALG has a lower bound on the terminal wealth. The terminal wealth of the UP is bounded for all instances of $\mathbf{x}$ by:

Lemma 5.1 *Assume that the UP competes against a* <u>*finite*</u> *target set of arbitrary benchmark algorithms* $\mathfrak{B}$. *The UP divides the initial wealth* W_0 *into* Ω *equal parts and invests according to* B^{ω} ($B^{\omega} \in \mathfrak{B}$ *with* $\omega = 1, ..., \Omega$). *Then, the terminal wealth of the UP equals* $W_T(UP, \boldsymbol{x}) = \frac{1}{\Omega}\sum_{\omega=1}^{\Omega} W_T(B^{\omega}, \boldsymbol{x})$ *and the worst-case logarithmic wealth ratio is bounded as*

$$\begin{aligned}
\boldsymbol{W}_T(UP, \mathfrak{B}) &= \max_{\boldsymbol{x}} \max_{B^{\omega}\in\mathfrak{B}} \ln \frac{W_T(B^{\omega}, \boldsymbol{x})}{W_T(UP, \boldsymbol{x})} \\
&= \max_{\boldsymbol{x}} \ln \frac{\max_{B^{\omega}\in\mathfrak{B}} W_T(B^{\omega}, \boldsymbol{x})}{\frac{1}{\Omega}\sum_{\omega=1}^{\Omega} W_T(B^{\omega}, \boldsymbol{x})} \\
&\leq \max_{\boldsymbol{x}} \ln \frac{\max_{B^{\omega}\in\mathfrak{B}} W_T(B^{\omega}, \boldsymbol{x})}{\frac{1}{\Omega}\max_{B^{\omega}\in\mathfrak{B}} W_T(B^{\omega}, \boldsymbol{x})} \\
&= \ln\Omega
\end{aligned} \tag{5.11}$$

for all T on arbitrary $\boldsymbol{x}$ [compare with Cesa-Bianchi and Lugosi, 2006, p. 278].

In other words, consider the following. Given are Ω benchmark algorithms where each is executed in the same market $\mathbf{x}$ by an expert. An investor acts as the UP and puts $\frac{1}{\Omega}$ of the wealth on each benchmark ALG. In the worst-case scenario one asset of the market $\mathbf{x}$ remains the same and all other assets crashes $(m-1)$ (see also Kelly market in Section 3.5.5). Assume there are $\Omega = m$ benchmark algorithms, where each is investing the full wealth in one of the m assets. Then, in the worst-case an offline player chooses in each trading period the best benchmark ALG which consists of the one "winner" asset with the stagnant prices. In contrast, the online player acts as UP in the worst-case and diversifies over all benchmark algorithms. Its fraction in the one stagnant asset will be at least $\frac{1}{\Omega}$. In this case, starting with $W_0 = 1$ the offline player receives $W_T^* = 1$ and the online player $W_T = \frac{1}{\Omega}$. It is easy to see that $\ln \frac{W_T^*}{W_T} = \ln \frac{1}{\frac{1}{\Omega}} = \ln \Omega$.

Lemma 5.2 *Assume the UP competes against an infinite target set of arbitrary benchmark algorithms. Then, the worst-case logarithmic wealth ratio is bounded as*

$$\begin{aligned}
\boldsymbol{W}_T(UP,\mathfrak{B}) &= \max_{\boldsymbol{x}} \max_{B^\omega \in \mathfrak{B}} \ln \frac{W_T(B^\omega,\boldsymbol{x})}{W_T(UP,\boldsymbol{x})} \\
&\leq \ln \binom{T+m-1}{T} \\
&= \ln \binom{T+m-1}{m-1} \\
&= \ln \left(\frac{(T+m-1)}{m-1} \cdot \frac{(T+m-1)-1}{m-1-1} \cdots \frac{(T+m-1)-(m-1-1)}{1} \right) \\
&\leq \ln (T+1)^{(m-1)} \\
&= (m-1)\ln(T+1)
\end{aligned} \tag{5.12}$$

for all T on arbitrary $\boldsymbol{x}$ with $m \geq 2$ assets [see Cover and Ordentlich, 1996, p. 353, Theorem 1].

The idea behind this lemma is as follows. Consider an investor invests in line with the UP and competes against all possible experts. In addition, assume the special case that the experts invest all or nothing in one asset at each trading period. The experts are allowed to jump between the assets, i.e, switching sequences. Thus, each expert chooses one of the m assets at each trading period. In fact, the calculation of the number of unique switching sequences is a combinatorial problem with repetition and without regard of the order. Thus, the number of unique experts in the worst case is calculated by

$$\binom{T+m-1}{T} \tag{5.13}$$

with given T and m. This in turn is limited from above by $(T+1)^{(m-1)}$. For a better understanding consider the following example with $T=2$ and $m=3$. When using Equation (5.13), then six different experts exist (i.e., $\{11, 22, 33, (12 \text{ or } 21), (13 \text{ or } 31), (23 \text{ or } 32)\}$ wherein the first (second) digit represents the selected asset of each expert at $t=1$ ($t=2$)). Recall, in a Kelly market asset $i=1, ..., (m-1)$ asset crashes and assume asset $i=m$ does not. Thus, there exists always only one expert which chooses no crash assets during $t=1, ..., T$ (here it is $\{33\}$) and receives a wealth of $W_T^*=1$. All other experts receive a wealth of 0. Hence, because the UP is the average of all experts the wealth of the UP is bounded by $W_T \geq \frac{1}{(T+1)^{(m-1)}}$ for $W_0=1$.

5.1.3 Risk-adjusted Portfolio Selection Algorithm

Dochow et al. [2014] provide the RAPS which uses the minimum value m_t^ω and maximum value M_t^ω of the period wealth for each expert up to time t to determine $\mathbf{b}_{(t+1)}$. The RAPS is defined for a finite number of CR-experts by

$$b_{i(t+1)}(RAPS) = \frac{\sum_{\omega=1}^{\Omega} \mathfrak{b}_i^\omega \frac{m_t^\omega}{M_t^\omega}}{\sum_{\omega=1}^{\Omega} \frac{m_t^\omega}{M_t^\omega}} \tag{5.14}$$

and for an infinite number of CR-experts by

$$\mathbf{b}_{(t+1)}(RAPS) = \frac{\int \mathfrak{b} \frac{\min(W_0, W_1(\mathfrak{b},\mathbf{x}),...,W_t(\mathfrak{b},\mathbf{x}))}{\max(W_0, W_1(\mathfrak{b},\mathbf{x}),...,W_t(\mathfrak{b},\mathbf{x})))} d\mathfrak{b}}{\int \frac{\min(W_0, W_1(\mathfrak{b},\mathbf{x}),...,W_t(\mathfrak{b},\mathbf{x}))}{\max(W_0, W_1(\mathfrak{b},\mathbf{x}),...,W_t(\mathfrak{b},\mathbf{x})))} d\mathfrak{b}}. \tag{5.15}$$

The terminal wealth of the RAPS is bounded in a Kelly market:

Theorem 5.1 *If the RAPS competes against a finite target set of CR-experts and if the set* $\boldsymbol{X}$ *contains only Kelly markets, then the worst-case* *logarithmic wealth ratio is*

$$\boldsymbol{W}_T(RAPS, \mathfrak{B}, \boldsymbol{X}) = \boldsymbol{W}_T(UP, \mathfrak{B}, \boldsymbol{X}). \tag{5.16}$$

Proof 5.1 *In a Kelly market the assets in the portfolio never increase, i.e.,* $0 \leq x_{it} \leq 1$. *Assume each CR-expert is starting in trading period* $t = 1$ *with an initial wealth* $W_0 = 1$. *Hence, the minimum and maximum period wealth of each CR-expert in a Kelly market is always* $0 \leq W_t(\mathfrak{b}^\omega, \boldsymbol{x}) \leq 1$ *and*

$$\frac{m_t^\omega}{M_t^\omega} = \frac{W_t(\mathfrak{b}^\omega, \boldsymbol{x})}{W_0} = W_t(B^\omega, \boldsymbol{x}). \tag{5.17}$$

Thus,

$$\begin{aligned} W_t(RAPS, \boldsymbol{x}) &= \frac{1}{\Omega} \sum_{\omega=1}^{\Omega} \frac{m_t^\omega}{M_t^\omega} \\ &= \frac{1}{\Omega} \sum_{\omega=1}^{\Omega} W_t(\mathfrak{b}^\omega, \boldsymbol{x}). \\ &= W_t(UP, \boldsymbol{x}) \end{aligned} \tag{5.18}$$

for all $t = 1, ..., T$. *It follows that*

$$\begin{aligned} \boldsymbol{W}_T(RAPS, \mathfrak{B}, \boldsymbol{X}) &= \max_{\boldsymbol{x} \in \boldsymbol{X}} \max_{B^\omega \in \mathfrak{B}} \ln \frac{W_T(B^\omega, \boldsymbol{x})}{W_T(RAPS, \boldsymbol{x})} \\ &= \max_{\boldsymbol{x} \in \boldsymbol{X}} \ln \frac{\max_{B^\omega \in \mathfrak{B}} W_T(B^\omega, \boldsymbol{x})}{\frac{1}{\Omega} \sum_{\omega=1}^{\Omega} \frac{m_T^\omega}{M_T^\omega}} \\ &= \max_{\boldsymbol{x} \in \boldsymbol{X}} \ln \frac{\max_{B^\omega \in \mathfrak{B}} W_T(B^\omega, \boldsymbol{x})}{\frac{1}{\Omega} \sum_{\omega=1}^{\Omega} W_T(B^\omega, \boldsymbol{x})} \\ &\leq \max_{\boldsymbol{x} \in \boldsymbol{X}} \ln \frac{\max_{B^\omega \in \mathfrak{B}} W_T(B^\omega, \boldsymbol{x})}{\frac{1}{\Omega} \max_{B^\omega \in \mathfrak{B}} W_T(B^\omega, \boldsymbol{x})} \\ &= \ln \Omega \\ &= \boldsymbol{W}_T(UP, \mathfrak{B}, \boldsymbol{X}). \end{aligned} \tag{5.19}$$

□

Lemma 5.3 *Assume the UP and the RAPS compete against a finite target set of CR-experts, uniformly distributed from the simplex* $\mathfrak{B}_m$. *If* $\boldsymbol{x}$ *is a Kelly* *market, then always*

$$b_{i(t+1)}(RAPS) = b_{i(t+1)}(UP). \tag{5.20}$$

Proof 5.2 *Recall that in a Kelly market* $\frac{m_t^\omega}{M_t^\omega} = W_t(\mathfrak{b}^\omega, \boldsymbol{x})$. *Thus,*

$$\begin{aligned} b_{i(t+1)}(RAPS) &= \frac{\sum_{\omega=1}^{\Omega} \mathfrak{b}_i^\omega \frac{m_t^\omega}{M_t^\omega}}{\sum_{\omega=1}^{\Omega} \frac{m_t^\omega}{M_t^\omega}} \\ &= \frac{\sum_{\omega=1}^{\Omega} \mathfrak{b}_i^\omega W_t(\mathfrak{b}^\omega, \boldsymbol{x})}{\sum_{\omega=1}^{\Omega} W_t(\mathfrak{b}^\omega, \boldsymbol{x})} \\ &= b_{i(t+1)}(UP) \end{aligned} \tag{5.21}$$

□

The findings are also true for an infinite number of experts with arbitrary benchmark algorithms, cf. Lemma 5.2.

5.1.4 Combined Risk-adjusted Portfolio Selection Algorithm

The CRAPS combines the weighting mechanism of the experts based on the RAPS and the UP. The idea is to combine the risk measure $\frac{m_t^\omega}{M_t^\omega}$ with the measure for the return on investment $W_t(\mathfrak{b}^\omega, \mathbf{x})$ to produce a simple risk-adjusted performance measure.

The CRAPS is defined as

$$b_{i(t+1)}(CRAPS) = \frac{\sum_{\omega=1}^{\Omega} \mathfrak{b}_i^\omega \frac{m_t^\omega}{M_t^\omega} W_t(\mathfrak{b}^\omega, \mathbf{x})}{\sum_{\omega=1}^{\Omega} \frac{m_t^\omega}{M_t^\omega} W_t(\mathfrak{b}^\omega, \mathbf{x})} \tag{5.22}$$

if the number of experts are finite. For an infinite number of CR-experts it is

$$\mathbf{b}_{(t+1)}(CRAPS) = \frac{\int \mathfrak{b} \frac{\min(W_0, W_1(\mathfrak{b},\mathbf{x}),\ldots,W_t(\mathfrak{b},\mathbf{x}))}{\max(W_0, W_1(\mathfrak{b},\mathbf{x}),\ldots,W_t(\mathfrak{b},\mathbf{x})))} W_t(\mathfrak{b}, \mathbf{x}) d\mathfrak{b}}{\int \frac{\min(W_0, W_1(\mathfrak{b},\mathbf{x}),\ldots,W_t(\mathfrak{b},\mathbf{x}))}{\max(W_0, W_1(\mathfrak{b},\mathbf{x}),\ldots,W_t(\mathfrak{b},\mathbf{x})))} W_t(\mathfrak{b}, \mathbf{x}) d\mathfrak{b}}. \tag{5.23}$$

The terminal wealth of the CRAPS is bounded in a Kelly market:

Theorem 5.2 *If the CRAPS competes against a finite target set of CR-experts, uniformly distributed from the simplex $\mathfrak{B}_m$, then CRAPS is bounded from below by $\boldsymbol{W}_T(RAPS, \mathfrak{B}, \boldsymbol{X})$.*

Proof 5.3 *In a Kelly market the loss of the CRAPS is never greater than the loss of the RAPS because*

$$\frac{m_t^{\omega}}{M_t^{\omega}} \times W_t(\mathfrak{b}^{\omega}, \boldsymbol{x}) \leq \frac{m_t^{\omega}}{M_t^{\omega}} \tag{5.24}$$

is always true because $\frac{m_t^{\omega}}{M_t^{\omega}} = W_t(\mathfrak{b}^{\omega}, \boldsymbol{x})$. This can be rewritten as

$$\begin{aligned} W_t(\mathfrak{b}^{\omega}, \boldsymbol{x}) \times W_t(\mathfrak{b}^{\omega}, \boldsymbol{x}) &\leq W_t(\mathfrak{b}^{\omega}, \boldsymbol{x}) \\ (W_t(\mathfrak{b}^{\omega}, \boldsymbol{x}))^2 &\leq W_t(\mathfrak{b}^{\omega}, \boldsymbol{x}). \end{aligned} \tag{5.25}$$

Note that the left (right) side of the inequalities is the weighting term for the experts of CRAPS (RAPS). The last inequality is always true in a Kelly market because $W_t(\mathfrak{b}^{\omega}, \boldsymbol{x}) = [0, 1]$ for $t = 1, ..., T$.

□

The findings are also true for an infinite number of experts with arbitrary benchmark algorithms, cf. Lemma 5.2.

5.2 Comparison of Competitiveness

Consider the performance guarantee of an ALG with respect to W_T in comparison to BCR. This guarantee is done without any statistical assumptions. For this a worst-case logarithmic wealth ratio $\mathbf{W}_T(ALG, \mathfrak{B}_m, \mathbf{X})$ is determined. It guarantees to have at least $W_T(ALG, \mathbf{x}) \geq \frac{W_T(BCR, \mathbf{x})}{exp(\mathbf{W}_T(ALG, \mathfrak{B}_m, \mathbf{X}))}$ for all $\mathbf{x} \in \mathbf{X}$. The comparison of the RAPS and the CRAPS is limited to algorithms which are analyzed in the competitive framework, i.e., called universal algorithms or FTW algorithms. Table 5.1 gives an overview on the most prominent universal algorithms from the literature. In addition to the worst-case logarithmic wealth ratio the limiting behavior for increasing m and T are shown ($O(\cdot)$). In addition, the conditions on $\mathbf{x}$ are given in which the $\mathbf{W}_T(ALG, \mathfrak{B}_m, \mathbf{X})$ is true.

Due to the different design of each online ALG, the proofs for the worst-case logarithmic wealth ratio are carried out in consideration of various

ALG	Constraints on $\mathbf{x}$	$\mathbf{W}_T(ALG,\mathfrak{B}_m,\mathbf{X})$	$O(\mathbf{W}_T(ALG,\mathfrak{B}_m,\mathbf{X}))$
SCR	$\max_{t,\tau}\|\mathbf{x}_t-\mathbf{x}_\tau\|=X<\infty$	$\ln\left(\frac{1}{C}(T-1)^{2X^2/\delta}\right)$	$O(X^2\ln T)$
UP	$0\leq x_{it}<\infty$	$(m-1)\ln(T+1)$	$O(m\ln T)$
EG	$0<x^{min}\leq x_{it}\leq x^{max}<\infty$	$\frac{x^{max}}{x^{min}}\sqrt{\frac{T}{2}\ln m}$	$O(\sqrt{T\ln m})$
ONS	$0<x^{min}\leq x'_{it}\leq x^{max}=1$	$\frac{1}{\beta}m\ln\left(\frac{mT}{(x^{min})^2}\right)+\frac{\beta}{2}$	$O(m\ln(mT))$
RAPS	$0\leq x_{it}\leq 1$	$(m-1)\ln(T+1)$	$O(m\ln T)$
CRAPS	$0\leq x_{it}\leq 1$	$(m-1)\ln(T+1)$	$O(m\ln T)$

Table 5.1: Summary of worst-case logarithmic wealth ratio of selected universal algorithms plus the RAPS and the CRAPS

assumptions. Consider the following remarks on the proofs provided in the literature:

- SCR [Gaivoronski and Stella, 2000, pp. 177-179, Theorem 3]: Let δ and C be two given constants. From a simplified perspective, the extent of variation between $\mathbf{x}_t$ and $\mathbf{x}_\tau$ with $t,\tau = 1,...,T$ is quantified by the Euclidean distance. The provided worst-case logarithmic wealth ratio assumes that the maximum Euclidean distance is known in advance. This maximum distance is quantified by X.

- UP [Cesa-Bianchi and Lugosi, 2006, pp. 283-284, Theorem 10.3]: In fact, no assumptions regarding $\mathbf{x}$ are made. The constraints on $\mathbf{x}$ are due to the definition of the PSP.

- EG [Cesa-Bianchi and Lugosi, 2006, pp. 285-286, Theorem 10.4]: The EG is a parametric ALG with a learning parameter η. For the proof of the lower bound it is assumed that the return factors are restricted by positive constants of the minimum value and maximum value for x_{it}. Let x^{min} and x^{max} be these given constants. Then, for the proof $\eta = \frac{x^{min}}{x^{max}}\sqrt{\frac{8\ln m}{T}}$.

- ONS [Agarwal et al., 2006, p. 11, Theorem 1]: For the lower bound of the parametric ONS, all x_{it} are scaled such that the maximum value of x'_{it} is equal to $x^{max} = 1$. In addition, x^{min} must be greater 0, which is called no-junk-bond assumption [compare with Agarwal et al., 2006, p. 2]. It is assumed that the learning parameter $\eta = 0$ and the worst-case logarithmic wealth ratio depends on the input parameter β.

- RAPS [Dochow et al., 2014]: The proof for the worst-case logarithmic wealth ratio is based on the property that for all $x_{it} \leq 1$ the RAPS is equal to the UP. Thus, in this scenario they have both the same lower bound.

- CRAPS: In the worst-case scenario with $x_{it} \leq 1$, the CRAPS puts always less or equal weights to the "bad" assets than UP and RAPS.

Summing up, assume the investment horizon T is a more critical parameter than the number of assets m for an investor to achieve a $\mathbf{W}_T(ALG, \mathfrak{B}_m, \mathbf{X})$ as low as possible. In other words, m is much smaller than T. The limiting behavior of $\mathbf{W}_T(ALG, \mathfrak{B}_m, \mathbf{X})$ is given by $O\left(\mathbf{W}_T(ALG, \mathfrak{B}_m, \mathbf{X})\right)$. Thus, the bounds of the SCR, the EG and the ONS are not as tight as for the UP, the RAPS and the CRAPS. Following [Cesa-Bianchi and Lugosi, 2006, p. 284], the worst-case performance of UP with respect to W_T is basically not improvable.

5.3 Numerical Results

The assessment of an ALG exclusively with the competitive analysis leads to a very pessimistic view. The assessment of the practical applicability should also be done by a statistical analysis with the help of numerical experiments. In line with former works the considered algorithms are empirically evaluated on the standard dataset[7] with NYSE data, which is most commonly used by the literature, e.g., Cover [1991] and Borodin et al. [2004].

The aim is, to answer the question whether or not the RAPS and the CRAPS are able to outperform some specific benchmarks in terms of return on investment, risk and risk-adjusted performance. The following algorithms are considered as benchmarks:

(i) **Online**: The online algorithms are limited to algorithms with a performance guarantee, i.e., FTW algorithms. All of them start in $t = 1$ with a naive diversification. To evaluate the effectiveness of experts algorithms, the SCR and the UP are considered. The minimum distance between the allocations of two CR-experts is set to $a = 0.01$ such that $\Omega = 101$ CR-experts in a market with $m = 2$ are taken into account. In addition, the influence of first and second order information is investigated by using the algorithms EG with $\eta = 0.05$ and the ONS with $\beta = 1$ and $\delta = 0.125$. (For further explanations concerning the algorithms read Section 4.2)

[7]The dataset from which the markets are generated can be found here: `http://www.cs.bme.hu/~oti/portfolio/data/nyseold.zip`

(ii) **Offline**: As typical offline benchmarks the BA and the BCR is used to quantify the maximum return on investment on the considered markets. Since the RAPS and the CRAPS incorporate risk management the offline benchmark BVCR is examined. BVCR is showing the performance of the CR-expert with the lowest variance at the end of the investment horizon (see Section 3.3).

(iii) **Non-Learning**: To incorporate an ALG which is not affected by the market sequence $\mathbf{x}$ itself, the UCR is employed (see Section 3.3.3).

The standard dataset with NYSE data includes daily price relatives of 36 assets in the range from 07/03/1962 to 12/31/1984 ($T = 5651$;≈22.60 years). The considered markets are limited to $m = 2$. Thus, $\binom{36}{2} = 630$ unique markets with $T = 5651$ are generated. For all 630 markets RAPS, CRAPS and the described benchmark algorithms are run. The following performance measures as described in Section 3.2 are calculated with 250 trading days in a year: (i) The terminal wealth (W_T), (ii) the *average annual percent yield* (APY), the *annual standard deviation* (ASTDV) and the *reward-to-variability ratio* (RVR). The intial wealth is set to $W_0 = 1$.

In addition, it is limited to four prominent pairs of assets from the literature ({comme and kinar, iro and kinar, coke and ibm, comme and meico}). This makes the findings comparable with existing results from the literature, cf. [Cover, 1991], [Helmbold et al., 1998], [Gaivoronski and Stella, 2000] and [Agarwal et al., 2006]. The APY, ASTDV and the correlation between return factors of the two assets are illustrated. The first asset mentioned corresponds to the *best asset portfolio* (BA) in the particular market. The following market constellations are considered:

- Market #1: **comme and kinar** (W_T: 52.02 and 4.13; $ASTDV$: 0.395 and 0.791; Correlation: 0.064; Volatile and stagnant, uncorrelated)
- Market #2: **iro and kinar** (W_T: 8.92 and 4.13; $ASTDV$: 0.538 and 0.791; Correlation: 0.041; Volatile, uncorrelated)
- Market #3: **coke and ibm** (W_T: 13.36 and 12.21; $ASTDV$: 0.206 and 0.221; Correlation: 0.388; Non-volatile, highly correlated)
- Market #4: **comme and meico** (W_T: 52.02 and 22.92; $ASTDV$: 0.395 and 0.490; Correlation: 0.067; Volatile, uncorrelated)
- Market #Avg: When focusing on the average performance of an ALG. Let $Perf(ALG, \mathbf{x})$ be a performance measure as $W_T, APY, ASTDV$ and RVR.

Consider the arithmetic mean of $Perf(ALG, \mathbf{x})$ over the 630 markets as a performance value for ALG which one can expect, on average, when choosing randomly one of the considered markets, i.e., $\mathbf{E}[Perf(ALG, \mathbf{x})]$.

The performance of an ALG must be evaluated by taking the risk and the corresponding risk-adjusted performance into account. Larger return on investment can be generated if one receives systematic higher risk [Modigliani and Modigliani, 1997]. The two algorithms RAPS and CRAPS are created by the modification of UP by taking risk management into account. A risk measure quantifies the risk of each expert in terms of the maximum fluctuation of the period wealth up to time t. This is why RVR is used to evaluate the performance of RAPS and CRAPS in comparison to the benchmark algorithms. Normally, the RVR requires a risk-free percentage yield R_f. For reasons of simplification, it is set to $R_f = 0\%$.

In Table 5.2 the empirical results are presented. The following conclusions for RAPS and CRAPS in comparison to the specific benchmark ALG can be drawn.

- BCR: RAPS is able to beat BCR in some market instances in terms of W_T (or APY). CRAPS is not able to do this. In general, RAPS and CRAPS trade with lower risk. This results on average into a higher RVR of RAPS and CRAPS in comparison to BCR.

- BA: In general, RAPS and CRAPS clearly outperform BA for all considered performance measures on #1-#4.

- BVCR: On the considered markets and on average, RAPS dominates BVCR in terms of W_T (or APY). In contrast, CRAPS does not dominate on the considered markets, but on average. As expected, the ASTDV of BVCR is in general lower or equal than these of RAPS and CRAPS. However, the results for RVR is not consistent. But on average, RAPS and CRAPS promise a larger RVR than BVCR.

- UCR: RAPS is able to beat UCR in all considered performance measures. In contrast, CRAPS is able to beat UCR only in terms of ASTDV.

- UP: RAPS and CRAPS achieve always a larger W_T (or APY) than UP. This result is remarkable because UP has a more general lower bound on W_T. In markets with larger volatility of assets RAPS and CRAPS generate a larger W_T than UP. RAPS is not able to generate always a lower ASTDV than UP on the four markets, but on average it does. In contrast, CRAPS has a lower ASTDV for the considered markets and on the average.

Market	$m = 2$	#1	#2	#3	#4	#Avg
BCR	W_T	144.00	73.70	15.07	102.96	26.57
	APY	0.25	0.21	0.13	0.23	0.16
	$ASDTV$	0.39	0.48	0.18	0.32	0.26
	RVR	0.63	0.44	0.70	0.71	0.61
BA	W_T	52.02	8.92	13.36	52.02	20.72
	APY	0.19	0.10	0.12	0.19	0.14
	$ASTDV$	0.40	0.54	0.22	0.40	0.29
	RVR	0.48	0.19	0.55	0.48	0.50
BVCR	W_T	116.81	58.85	14.95	102.96	17.37
	APY	0.23	0.20	0.13	0.23	0.12
	$ASTDV$	0.37	0.46	0.18	0.32	0.20
	RVR	0.64	0.43	0.71	0.71	0.63
UCR	W_T	118.69	72.58	15.02	98.89	21.84
	APY	0.24	0.21	0.13	0.23	0.15
	$ASTDV$	0.46	0.49	0.18	0.33	0.23
	RVR	0.52	0.43	0.71	0.69	0.64
SCR	W_T	26.36	16.56	5.48	28.14	12.13
	APY	0.16	0.13	0.08	0.16	0.12
	$ASTDV$	0.46	0.55	0.21	0.34	0.27
	RVR	0.34	0.24	0.38	0.46	0.44
UP	W_T	80.54	39.97	14.24	74.08	18.89
	APY	0.21	0.18	0.12	0.21	0.14
	$ASTDV$	0.43	0.49	0.18	0.31	0.23
	RVR	0.50	0.36	0.69	0.67	0.61
EG	W_T	110.96	64.43	14.90	94.28	21.29
	APY	0.23	0.20	0.13	0.22	0.14
	$ASTDV$	0.45	0.49	0.18	0.32	0.23
	RVR	0.52	0.41	0.70	0.69	0.64
ONS	W_T	357.22	28.49	18.33	85.89	27.21
	APY	0.30	0.16	0.14	0.22	0.16
	$ASTDV$	0.38	0.54	0.18	0.45	0.24
	RVR	0.77	0.30	0.76	0.48	0.65
RAPS	W_T	127.96	91.33	15.38	109.57	23.07
	APY	0.24	0.22	0.13	0.23	0.15
	$ASTDV$	0.46	0.48	0.18	0.32	0.22
	RVR	0.52	0.46	0.71	0.71	0.68
CRAPS	W_T	93.08	49.64	14.57	86.05	20.08
	APY	0.22	0.19	0.13	0.22	0.14
	$ASTDV$	0.42	0.48	0.18	0.31	0.22
	RVR	0.53	0.39	0.70	0.70	0.65

Table 5.2: Empirical comparison of algorithms for various performance meausres.

When focusing on RVR, RAPS and CRAPS always dominate UP. When considering the ASTDV of UP in comparison to RAPS (CRAPS), it is smaller in 393 (500) of the 630 considered markets, i.e., in 62.4% (79.4%) of the markets the ASTDV is smaller.

- EG: Only the RAPS dominates EG in terms of APY and RVR. The results for APY and RVR of CRAPS are not clear. The results of RAPS and CRAPS in terms of ASTDV are slightly less than EG.
- SCR: RAPS and CRAPS dominate SCR in all terms of performance measures on all considered markets and on average.
- ONS: The results of ONS seem to vary greatly. On average, the APY of ONS in comparison to RAPS and CRAPS is larger. But, also the risk in terms of ASTDV is larger. Hence, the RVR of RAPS is larger than that of ONS, but the RVR of CRAPS is equal to that of ONS.

In addition, regarding the characteristics of the considered markets with $m = 2$, it can be found for W_T (or APY) that:

- In markets with volatile and uncorrelated assets large gains in comparison to BA can be generated. When focusing on the average results, it can be seen that ONS is better than all other benchmark algorithms. RAPS and CRAPS are second and third best of all considered algorithms. Respectively, RAPS outperforms BCR in two cases (#2 and #4). The performance of ONS, RAPS and CRAPS increases with the market volatility whereas the W_T (or APY) of the other algorithms decreases (see #1).
- In markets with non-volatile and highly correlated assets, rebalancing does not help to generate "extra" return on investment. In #3 almost all algorithms (except SCR) achieve an APY of around 13%. This is as good as BCR and BVCR and a little higher than BA. RAPS and ONS are the only algorithms which are able to outperform BCR.

In general, RAPS has the best RVR on average. It follows CRAPS and ONS. Note that SCR is the overall worst algorithm in terms of APY and RVR. In Figure 5.1 the average results of APY and ASTDV are illustrated, i.e., $\mathbf{E}[APY]$ in relation to $\mathbf{E}[ASTDV]$. The slope of a theoretical line between the zero point and each spot of an algorithm in the figure indicates the extent of the RVR. The lower the slope the higher the RVR. Summing up, the empirical results suggest that the application of RAPS let expect a large trade-off between return on investment and risk, i.e., risk-adjusted performance.

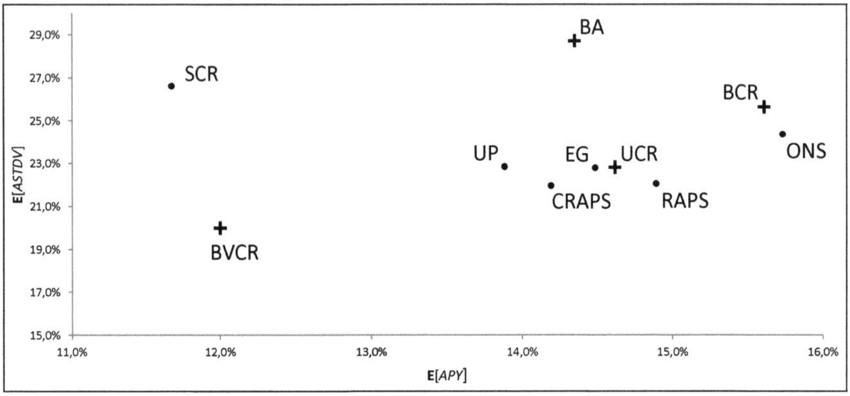

Figure 5.1: $\mathbf{E}[APY]$ in relation to $\mathbf{E}[ASTDV]$ for the considered algorithms with 630 different markets

5.4 Conclusions

In line with [Borodin and El-Yaniv, 1998, pp. 309-311] and [Li and Hoi, 2014, p. 29], existing online algorithms in the MLC do not consider risk management. In this chapter the UP of Cover [1991] is modified such that risk management is incorporated. As a result, two new algorithms (RAPS and CRAPS) are proposed and discussed without making any statistical assumption for a risky market. The lower bounds of RAPS and CRAPS are proofed in a worst-case scenario with respect to the terminal wealth. The bounds give a performance guarantee for an investor in a Kelly market. In comparison with other guaranteeing algorithms the bounds of RAPS and CRAPS are as tight as for UP, which is basically not improvable.

Empirical results show that especially RAPS is encouraging in terms of risk-adjusted performance measured by RVR. Although CRAPS is able to reduce risk, but on average RAPS can generate more return on investment with the same risk. Compared to the considered benchmark algorithms, the results of RAPS are promising. The significance of the empirical results is still an open issue. In addition, as UP the time complexities of RAPS and CRAPS are exponential when the number of assets m in the market increases (see also Lemma A.3 in the appendix). Thus, future work focus on the development of an ALG which incorporates risk management but with lower time complexity.

6 Empirical Testing of Algorithms

In this chapter all considered online algorithms of the previous chapters are simulated and tested on a large number of various market instances. As benchmark algorithms the uniform buy-and-hold portfolio and the uniform constant rebalancing portfolio are applied. The algorithms are tested multiple times with respect to performance measures of return on investment, risk and risk-adjusted performance.

6.1 Preliminaries

The objective of this chapter is to investigate the empirical performance of online algorithms for the PSP, as described in the previous chapters, based on a statistical analysis. One object of research is the average performance with respect to return on investment, risk and risk-adjusted performance (cf. Section 3.2). The ability of the online algorithms to beat the benchmark algorithms (i) UBH and (ii) UCR is another object of research. UBH and UCR are employed because as non-learning benchmarks their portfolio allocations are independent of the concrete market instance, but showing promising results in previous empirical investigations and are often used in practice by investors [see DeMiguel et al., 2009]. Offline benchmarks are not taken into account because an investor is not able to apply such benchmarks without having future information.

Previous investigations for the empirical performance of online algorithms claim that FTL algorithms show much better return on investment than FTW algorithms, e.g., [Li et al., 2013]. To conduct a new empirical investigation a new dataset with 30-years prices of 240 NYSE stocks in the range between 1984 and 2014 from the website `http://finance.yahoo.com/` is obtained.

The chapter is structured as follows: First, the investigated online algorithms are introduced. Then, associated related work is provided. Afterwards, a brief description of the assets in the dataset is given and the two test designs to be carried out are explained. It follows the numerical results of the average empirical performance and the empirical investigation of the ability to beat UBH and UCR. The chapter concludes with final remarks.

6.1.1 Algorithms and Parameters

The following online algorithms are tested with the specified parameters from literature (for further explanations of the algorithms see Chapter 4 and 5):

1. SCR - Successive Constant Rebalanced Portfolio Algorithm: The portfolio allocation of SCR is calculated based on a number of Ω experts with daily constant rebalancing. Thus, the larger Ω the more precise SCR performs. When considering markets with $m = 2$ assets then $\Omega = 21$ experts are applied, which gives a precision of $a = 5\%$. When markets with $m = 5$ ($m = 10$) are considered the portfolio allocation of the ALG is estimated with $\Omega = 100$ ($\Omega = 200$) random experts (for more details see Section 3.3).

2. UP - Universal Portfolio Algorithm: As for SCR the same Ω experts are employed (see also [Cover, 1991, pp. 18-28] for $m = 2$).

3. EG - Exponential Gradient Algorithm: Helmbold et al. [1998, p. 338] propose a learning rate of $\eta = 0.05$.

4. ONS - Online Newton Step: Agarwal et al. [2006, p. 13] propose as parameters $\beta = 1$ and $\delta = 0.125$.

5. AC - Anti Correlation Algorithm: From empirical tests of Borodin et al. [see Figure 2 2004, p. 586], it can be concluded that AC with a window size of $w = 5$ achieves encouraging results.

6. PAMR - Passive Aggressive Mean Reversion Algorithm: It is limited to PAMR1 with a sensitivity parameter $\hat{\bar{x}} = 0.5$. It is in line with Li et al. [2012, p. 239].

7. CWMR - Confidence Weighted Mean Reversion Algorithm: It is limited to CWMRV. According to Li et al. [2013, p. 438], let $\hat{\bar{x}} = \exp(-0.5) = 0.61$ and let the confidence parameter $\phi = 2$.

8. OLMAR - Online Moving Average Mean Reversion Algorithm: In line with Li and Hoi [2012, p. 6], a window size $w = 5$ and a mean reversion parameter $\hat{\bar{x}} = 10$ are taken.

9. RMR - Robust Median Mean Reversion Algorithm: Referring to Huang et al. [2013, p. 2010], the window size $w = 5$, the mean reversion parameter $\hat{\hat{x}} = 5$ and the number of iterations $K = 200$ are employed. A proper value for the toleration level $\hat{\tau}$ is missing in the literature. Here it is set to $\hat{\tau} = 1$.

10. RAPS - Risk-Adjusted Portfolio Selection Algorithm: As for SCR and UP, the same experts are employed (for more details see [Dochow et al., 2014] and Chapter 5).

11. CRAPS - Combined Risk-Adjusted Portfolio Selection Algorithm: According to SCR, UP and RAPS, the same experts are taken (for more details see Chapter 5).

Recall, concerning the eleven investigated online algorithms, four are FTW algorithms (SCR, UP, EG and ONS), five are FTL algorithms (AC, PAMR, CWMR, OLMAR and RMR) and two are risk algorithms (RAPS and CRAPS).

6.1.2 Related Work

To evaluate the performance of an online ALG for the PSP based on a statistical analysis several test designs exist in the literature. The following components of a test design are identified in [Cover, 1991], [Helmbold et al., 1998], [Gaivoronski and Stella, 2000], [Borodin et al., 2004], [Agarwal et al., 2006], [Li and Hoi, 2012], [Li et al., 2012], [Li et al., 2013], [Huang et al., 2013] and [Dochow et al., 2014] to make the included performance evaluations comparable:

1. **Algorithm(s)**: The algorithm(s) to be tested is (are) clear in advance. Additionally, benchmarks are defined.

2. **Object of Research**: Each test design is based on an underlying research question.

3. **Dataset(s)**: At least one dataset with sufficient data (assets and asset prices) is required.

4. **Sampling Technique**: How to handle one or more datasets to generate market instances depend on the properties of the underlying dataset(s) and the object of research.

5. **Performance Measure**: The final evaluation depends on a measure for the return on investment, risk or risk-adjusted performance. Moreover, this can be combined with statistical measures or hypothesis testing.

Within the considered literature four categories of the object of research are identified:

- #a - Empirical Performance: A concrete ALG is applied on selected market instances to show its exemplary performance. The performance is compared with those of other algorithms. Most prominent examples are market instances with $m = 2$ assets based on a NYSE dataset with 36 assets (see also Table 6.1), e.g., [Cover, 1991], [Helmbold et al., 1998] and [Gaivoronski and Stella, 2000].
- #b - Algorithm Parameters: The impact of different values for the parameters on a performance measure for one concrete parametric ALG is analyzed. For example, Helmbold et al. [1998] investigate a parametric ALG for different values of a learning parameter. In addition, Borodin et al. [2004] vary the value of a window size parameter.
- #c - Market Properties: The performance of an ALG varies with different market properties. The impact of the variation in the market properties (e.g., number or riskiness of assets in the market) on a concrete performance measure is investigated. For instance, Agarwal et al. [2006] compare the return on investment of an online ALG between low and high volatile markets.
- #d - Significance of Outperformance: To investigate whether "good" performance of an ALG is simply based on luck hypothesis testing is executed. Examples are given in [Li et al., 2012] and [Li and Hoi, 2012] where the return on investment of an ALG is compared with the return on investment of the market (more precisely: UBH) based on the statistical test OTT (for more details see Section 3.4.2).

The associated datasets with daily data, which have been employed most for empirical statistical analysis, can be found in Table 6.1. They are *New York Stock Exchange dataset* (NYSE), *Toronto Stock Exchange dataset* (TSX), *Standard & Poors 500 dataset* (SP500), *Dow Jones Industrial Average dataset* (DJIA), *New York Stock Exchange dataset 2* (NYSE2) and *Morgan Stanley Capital International dataset* (MSCI). It is noticeable, the number of assets in the datasets, with one exception, is quite small. TSX includes 88 assets but only daily data for five years. Larger datasets would be desirable in order to generate more robust statements concerning the empirical performance of online algorithms.

#	Name	Time frame	Trading periods (Assets)	Introduced by
1	NYSE	07/03/1962 – 12/31/1984	5651 (36)	[Cover, 1991]
2	TSX	01/04/1994 – 12/31/1998	1259 (88)	[Borodin et al., 2004]
3	SP500	01/02/1998 – 01/31/2003	1276 (25)	[Borodin et al., 2004]
4	DJIA	01/14/2001 – 01/14/2003	507 (30)	[Borodin et al., 2004]
5	NYSE2	01/01/1985 – 06/30/2010	6431 (23)	[Li et al., 2012]
6	MSCI	04/01/2006 – 03/31/2010	1043 (24)	[Li et al., 2012]

Table 6.1: Summary of selected datasets with daily data used in the literature

Based on the employed datasets, the following three main sampling techniques are determined to produce a variation of numerical results. A brief description of these sampling techniques and some critics are given as follows:

- #1 - Simplified Market: A specific number of assets from a dataset is taken intentionally or randomly. This is done in order to get a first impression of the empirical performance of an ALG. By changing the number of assets in the market, the influence of this number on the performance is examined. In earlier studies this number is set to $m = 2$ and $m = 3$, e.g., [Cover, 1991] and [Helmbold et al., 1998]. This is probably because in the past the computational capacity is not sufficient to calculate markets with larger m. Moreover, it is conceivable that asset combinations are chosen in which the concrete ALG in terms of one specific performance measure worked well.

- #2 - Full Market: The full dataset is used as one entire market instance. To be more realistic, an investor usually selects from a large set of assets. When the ALG faces a large number of assets m then a suitable ALG selects the "appropriate" asset(s) and builds a portfolio which promises a "good" performance. To be critical, if there are one or two "abnormal" assets within the dataset, then this can lead to doubtful conclusions. For example, Li et al. [2012, p. 243] test an ALG on the NYSE dataset with and without the asset "kinar". As a result, the ALG shows substantial different performance on the original and modified dataset.

- #3 - Inverse Full Market: The full dataset is used as one entire market instance and each return factor is replaced by $x'_{it} = \frac{1}{x_{it}}$ with $i = 1, ..., m$ and $t = 1, ..., T$. The motivation behind this approach is to produce a larger variety of market instances. The properties of the original full market with a larger (lower) number of increasing assets is reversed to a modified market with a lower (larger) number of increasing assets, e.g., [Borodin et al., 2004, p. 589].

In general, when a new online ALG is introduced in the literature, then its performance is compared with those ones of all previous state-of-the-art online algorithms (for more details see also Chapter 4) by an empirical investigation. Table 6.2 gives a summary of test designs which are applied in the associated publications where the original online ALG (see the first column) is introduced.

Algorithm (introduced)	Return on investment	Risk	Risk-adjusted performance	Sampling technique	Object of research	Dataset
UP	W_T	-	-	#1	#a	NYSE
EG	W_T, APY	-	-	#1	#a, #b, #c	NYSE
SCR	W_T	-	-	#1	#a	NYSE
AC	W_T, APY	ASTDV	RVR	#2, #3	#a, #b	NYSE, TSX, SP500, DJIA
ONS	W_T, APY	-	RVR	#1	#a, #c	NYSE
PAMR	W_T, APY	MDD	RVR, DDR	#2	#a, #c, #d	NYSE, NYSE2, TSX, SP500, DJIA, MSCI
CWMR	W_T, APY	ASTDV, MDD	RVR, DDR	#2	#a, #d	NYSE, NYSE2, TSX, MSCI
OLMAR	W_T	-	-	#2	#a, #d	NYSE, NYSE2, TSX
RMR	W_T, APY	-	-	#2	#a, #d	NYSE, NYSE2, DJIA, MSCI
RAPS	W_T	-	-	#1	#a	NYSE

Table 6.2: Summary of test designs, which are used in the literature when the concrete online ALG is introduced. The algorithms are provided in chronological order based on the year of publication.

Concerning the performance evaluation based on a statistical analysis for each online ALG the following findings of the literature in the MLC are obtained:

- UP: Assets with low risk and a high correlation of the prices are less advantageous for online portfolio selection algorithms [Cover, 1991, p. 23].
- EG: The larger m in the market the greater is the achievable terminal wealth of EG. Empirically, EG performs better than UP, although UP is theoretically superior to EG [Helmbold et al., 1998, pp. 344,345].
- SCR: Gaivoronski and Stella [2000, p. 181] show empirically, a modified SCR is superior to UP and EG in terms of terminal wealth. A performance evaluation of the original SCR itself is missing.
- AC: It is the first ALG which exploits the mean reversion property of assets, i.e., FTL algorithm. AC has a striking outperformance in comparison to UBH and BA in terms of terminal wealth. But, there are some market instances where the terminal wealth of AC is far lower than the ones of the benchmarks. AC is more risky than the previously introduced online algorithms. UP and EG are not substantial beneficial in comparison to UCR [Borodin et al., 2004, p. 589].
- ONS: The larger m in the market the greater is the achievable terminal wealth of ONS. In addition, markets with high risky assets are beneficial for ONS. An investigation of the risk-adjusted performance show, the ONS achieves the largest RVR in comparison to BCR, UCR, UP and EG [Agarwal et al., 2006, pp. 13-15].
- PAMR: PAMR is significantly better than all other previous algorithms (including BCR) in terms of terminal wealth. This is due to the powerful exploitation of mean reversion for high risky assets when m is large. The excessive performance of PAMR in terms of terminal wealth is evaluated with hypothesis testing and can be therefore not based merely on luck. PAMR is a very risky ALG, but the risk-adjusted performance is still sufficient to compete with the benchmarks [Li et al., 2012, pp. 243-245].
- CWMR: The results of Li et al. [2013] for CWMR are similar to the results of PAMR but with a little improvement of the terminal wealth on the considered market instances.
- OLMAR: Li and Hoi [2012] demonstrate, although the terminal wealth of CWMR and PAMR are already excessive high on the used datasets, OLMAR provides an even larger terminal wealth.

- RMR: The RMR is a modification of the OLMAR and achieves again a larger terminal wealth than all previous algorithms on the same datasets [Huang et al., 2013].
- RAPS: Dochow et al. [2014] provide the first online ALG with risk management of the MLC. It is shown that RAPS is able to beat UCR, BCR and all other FTW algorithms for selected markets in terms of terminal wealth.

Summing up, each of the considered test designs from the literature has been created for the purpose of performance evaluation of one concrete online ALG. Other algorithms are exclusively used as benchmarks. Critically regarded, there is evidence that test designs have been chosen such that the results support the "positive" performance of the concrete introduced online ALG. In line with [Lo and MacKinlay, 1990, pp. 461-465] and Hsu and Kuan [2005, p. 608] such a problem is denoted as data snooping and is solved by applying additional datasets or sub-samples of a large dataset.

Hence, a comprehensive statistical analysis of all considered online algorithms in a whole is desired. This should be based on (i) a large dataset and (ii) the application of sub-samples [motivated also from Alpaydin, 2010, chap. 19].

6.1.3 Dataset and Description

Based on the availability of a large dataset for historical stock prices, data is collected from the website `http://finance.yahoo.com/`[8]. The most data is available for the New York Stock Exchange. All available stocks from this stock exchange on this website with at least prices from 01/01/1984 to 12/31/2014 are taken. Dividends are reinvested and stock splits are taken into account. After the selection, each asset is manually checked for unusual price jumps and if any exist, then the asset is excluded from the dataset. Finally, 240 stocks with 30-years price data between the years 1984 and 2014 remained (7564 trading days). It should be noted, despite the greatest care, the data may still contain errors. Nevertheless, the dataset should be large enough for robust testing of the considered online algorithms. Thus, the dataset is a collection of asset prices for the following 240 stock symbols:

AA, AAN, ABM, ABT, ADM, ADX, AEM, AEP, AET, AFG, AFL, AIG, AIR, AIT, AJG, ALE, ALK, ALX, AMD, AME, AON, AOS, AP, APA,

[8] Date of request: 01/07/2015

APD, ARW, ASA, ASH, ATO, ATW, AVP, AVT, AVY, AXE, AXP, AXR, AZZ, B, BA, BAM, BAX, BC, BCE, BCR, BDX, BEN, BF-B, BGG, BIO, BK, BKH, BLL, BMI, BMS, BMY, BOH, BP, BRK-A, BRT, BT, BWS, C, CAG, CAJ, CAS, CAT, CB, CBT, CCK, CHE, CL, CLX, CMI, CMS, CNP, CNW, COO, COP, CP, CSC, CSX, CVS, CVX, D, DBD, DCO, DD, DE, DIS, DNB, DOW, DTE, DUK, ED, EDE, EGN, EIX, EMR, ENZ, ESL, ETN, ETR, EXC, F, FAC, FCE-A, FDX, FL, FRM, FSS, FUR, GAS, GD, GE, GFF, GIS, GLT, GLW, GPC, GTY, GWW, GY, HAL, HD, HES, HMC, HON, HP, HPQ, HRG, HRS, HST, HUB-B, HUM, IBM, IFF, IP, JCP, JNJ, JPM, K, KMB, KO, KR, LLY, LMT, LNC, LPX, LUV, LXU, MAS, MCD, MDC, MDR, MDT, MEG, MMM, MO, MRK, MSI, MUR, MUX, NAV, NBL, NC, NEE, NEM, NI, NKE, NL, NOC, NSC, NU, NUE, NVO, NWL, NYT, OXY, PBI, PCG, PEG, PEI, PEP, PFE, PG, PHI, PKD, PKI, PKY, PNM, PNR, PNW, PPG, PSA, R, RAD, RBC, RDC, ROG, ROK, RSH, RTN, S, SF, SFY, SJW, SLB, SNE, SO, SSL, SYY, T, TAP, TE, TEN, TGT, THC, TM, TPC, TSO, TXT, TY, UIS, UNP, UTX, VLO, VZ, WEC, WFC, WFT, WGO, WMB, WMT, WRB, WSO, WWW, WY, XOM, XRX, Y

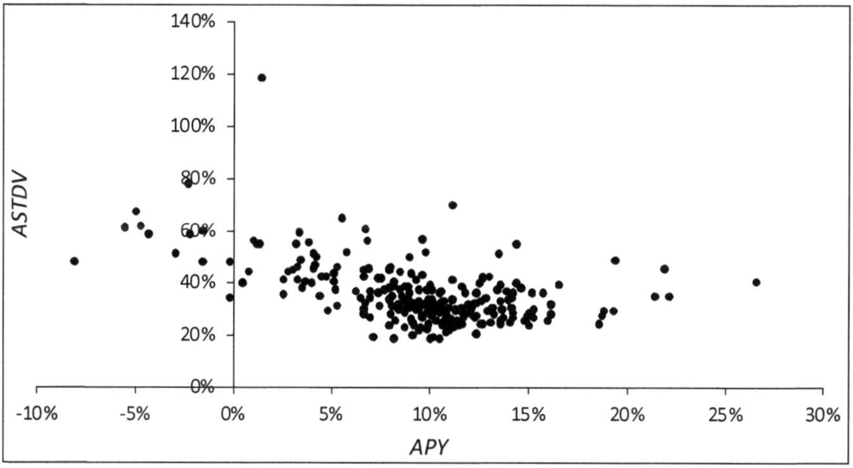

Figure 6.1: *APY* in relation to *ASTDV* for each asset of the dataset with a single asset buy-and-hold strategy

To illustrate the characteristics of the 240 assets, consider each asset as a single asset BH strategy. Then, the APY and the ASTDV with 250 trading days in a year for all strategies are calculated and illustrated in Figure 6.1.

The average of all APYs is 9.4% and the average of all ASTDVs is 35.3%. 12 of the 240 assets have an APY lower than 0% and 135 are larger than the average. 98 of the 240 assets have a larger ASTDV than the average, where MUX has an extraordinary risk with $ASTDV = 117.7\%$. All assets in the dataset are considered as risky assets.

6.2 Test Design

The following investigations contribute to the practical applicability of the considered online algorithms. The focus is only on the potential of the performance for the respective algorithms. The algorithms are analyzed from various performance perspectives because in practice there are different investors with varying purposes. The online algorithms execute a daily rebalancing and for simplification transaction costs and taxes are ignored. Thus, the results obtained can be interpreted as an indication for an upper bound of the performance on this dataset for each online ALG. It is assumed, if transaction costs are added plus executing a non-daily rebalancing, then the performance of each ALG gets worse. The extent may be different for each ALG. Nevertheless, consider the following simple classification of an investor: (i) A "risk-loving" investor may ignore the risk and is searching for an ALG which is promising the largest increase in wealth. (ii) In contrast, a risk averse investor avoids the risk of loss and is searching for an ALG that aims to obtain the highest possible retention of wealth at the end of the investment horizon. (iii) Beyond that, there is an investor who searches for the ALG with the largest expected increase in wealth in relation to the potential of loss in wealth.

In order to satisfy an investor as described in (i) the arithmetic mean of the APY on a large number of different markets is determined, i.e., *average annual percentage yield* ($\mathbf{E}[APY]$). Since the arithmetic mean is sensitive to outliers, in addition, the median of observations for the APY is calculated and so called *median annual percentage yield* ($\mathbf{MED}[APY]$). Referring to an investor of (ii), the *average annual standard deviation* ($\mathbf{E}[ASTDV]$) and the *average maximum drawdown* ($\mathbf{E}[MDD]$) are considered. An investor of (iii) is encountered through a simplified performance measurement of the *average risk-to-reward ratio* ($\frac{\mathbf{E}[APY]}{\mathbf{E}[ASTDV]}$) and the *average drawdown ratio* ($\frac{\mathbf{E}[APY]}{\mathbf{E}[MDD]}$). Summing up, the stated statistical measures are used to compare each online ALG with other online algorithms. This is done in order to indicate in practice whether the one ALG is promising a larger return on investment, lower risk or higher risk-adjusted performance than the others (for more details of the performance evaluation see Chapter 3).

The following steps are processed and describe the first empirical test design:

1. **Set Number of Assets**: Set m as number of assets in the investigated market. Here, to consider the effect on the algorithms when the number of assets in the market is varied, different values are regarded, i.e., $m = 2$, $m = 5$ and $m = 10$. However, all following steps are only valid for one specific m.

2. **Choose Random Assets**: Choose m assets randomly from the dataset. Note, no asset is selected more than once.

3. **Generate Market**: Take all historical prices ($T = 7564$) for the selected m assets and build the market as one random market instance.

4. **Set Number of Random Markets**: Set the desired number of random market instances N and repeat the previous two steps N times. In this test design it is always $N = 1000$ for all variations of m. Note that with $m = 2$ there are $\binom{240}{2}$ =2.87E+04 unique combinations to create a market with different asset compositions. For $m = 5$ there are 6.36E+09 and for $m = 10$ there are 1.44E+17. Thus, repeating markets are very unlikely.

5. **Run Algorithms and Benchmarks**: Run all considered algorithms and benchmarks on the N random market instances with the relevant algorithm parameters.

6. **Calculate Performance Measures**: Calculate the required performance measures for each ALG on each random market instance. More specifically, the APY, ASTDV and MDD are calculated with 250 trading days in a year.

7. **Calculate Statistics**: Calculate the statistical measures based on the performance measures. Precisely, it is $\mathbf{E}[APY]$, $\mathbf{MED}[APY]$, $\mathbf{E}[ASTDV]$ and $\mathbf{E}[MDD]$.

In addition, the results of such empirical test design when using one dataset is sometimes influenced by the dominance of a few assets or the involvement of a particularly negative (or positive) historical period of investigation. Nevertheless, the emergence of such "abnormal" properties in historical data is not discussed further here. However, a second test design is applied and is motivated to reduce the influence of such "abnormalities" on the performance evaluation of online algorithms, i.e., an investigation for the robustness of the results. It is based on sub-samples of the initial dataset. It is asked for the ability of an online ALG to beat a specific benchmark (UBH and UCR) for an explicit performance measure (APY, ASTDV and RVR).

The subsequent steps are done and describe the second empirical test design:

1. **Set Number of Assets**: Set the number of assets m, which should be included in the market instances when the algorithms are tested. This number is set to $m = 2$ and $m = 5$. Note, the following steps are only valid for one specific m.

2. **Set Number of Trading Periods**: Set the number of trading periods T, which should be included in each market instance when the algorithms are tested. This number is set to $T = 1250$, which is equivalent to a five-year trading investment horizon when one year consists of 250 trading days.

3. **Generate Sub-Datasets**: Separate the initial dataset into K non-overlapping sub-datasets which consists of an equal number of assets and trading periods. Note, the order of the initial trading periods is not changed. Basically, the dataset is only cut. The initial dataset, where the 240 assets are ordered by the asset name, is cut into 30 sub-datasets with 8 assets. Each of the 30 sub-datsets is cut again into three sub-datasets with each 2521 trading periods. At the end $K = 90$ sub-datasets exist where each consists of 8 unique assets and 2521 unique trading periods.

4. **Choose Random Assets**: Choose m assets randomly from each of the K sub-datasets. Note, no asset is selected more than once.

5. **Generate Market**: Take one random block of asset prices with $T = 1250$ trading periods for the selected m assets from each of the K sub-datasets. Thus, K random market instances with random assets and random historical starting time for the investment horizon are generated. Each market instance could have been chosen in the past from a real investor.

6. **Set Number of Random Markets**: Set the desired number of random market instances N and repeat the previous two steps N times. At the end, $K \times N$ random market instances exist. In this investigation, it is set to $N = 100$. Note, for $m = 2$ ($m = 5$) there exist $\binom{8}{2} = 28$ ($\binom{8}{5} = 56$) different combinations for the assets in the market. This is quite low. But, the market instances are also randomized by time. In each sub-dataset

$2521 - 1251 + 1 = 1271$ blocks with different (but overlapping) trading periods can be taken. Nevertheless, the probability of repeating market instances is low.

7. **Run Algorithms and Benchmarks**: Run all desired algorithms and benchmarks on the $K \times N$ random market instances with the relevant parameters for the algorithms.

8. **Calculated Performance Measures**: Calculate the required performance measures for each ALG on each random market instance. More specifically, the APY, ASTDV and RVR is calculated with 250 trading days in one year.

9. **Execute Hypothesis System**: Classify all resulting performance measures for all algorithms and benchmarks according to their origin with respect to the K sub-datasets. Execute K right-sided hypothesis systems with two samples ($H^{right-sided}$) for APY (and RVR) where the ALG is tested against one benchmark (UBH and UCR). Use a right-sided hypothesis system with two samples ($H^{left-sided}$) for ASTDV. Precisely, the PTWSRT is employed with $\alpha = 0.05$ (for more details see Section 3.4.2).

10. **Calculate Percentage of Outperformance**: Calculate the *percentage of outperformance* (POO) for each performance measure, showing the percentage of the ability of how an ALG is significantly able to beat a given benchmark on the K sub-datasets. A larger value indicates a higher potential of outperformance depending on the number of assets m, the specific performance measure (e.g., APY and ASTDV) and the investment horizon T. POO is calculated by

$$POO = \frac{1}{K} \sum_{k=1}^{K} H^{(k)} \tag{6.1}$$

where $H^{(k)}$ represents the output of the hypothesis system k which uses only data generated by sub-dataset k. Note, $H^{(k)} = 1$ if $H_1^{(k)}$ is accepted and otherwise $H^{(k)} = 0$ (see Section 3.4.2 for more details).

Summing up, two test designs are provided step by step. Note, historical findings should be treated with caution. The validity of such findings for the future depends on whether the future behaves similarly to the past.

6.3 Numerical Results: Expected Performance

Figure 6.2 shows the results of $\mathbf{E}[APY]$ and $\mathbf{MED}[APY]$ with various m (2, 5 and 10) on $N = 1000$ random market instances. The investment horizon is $T = 7564$ and the results are ordered by $\mathbf{E}[APY]$, starting with the smallest value. It is obvious, the APY of the FTL algorithms are much better than these for the remaining algorithms where OLMAR and RMR are always the best two. The distances between the performances increase with rising m. FTL algorithms work much better when they can select from a larger set of assets. Surprisingly, with $m = 10$ the $\mathbf{E}[APY]$ of the FTL algorithms is many times higher than that of the FTW algorithms. The $\mathbf{E}[APY]$ for all FTL algorithms are much higher than for $\mathbf{MED}[APY]$. This indicates, there are a few market instances where the FTL algorithms are able to generate a superior APY in comparison to the remaining market instances. This effect can be neglected for the FTW algorithms, the risk algorithms and the benchmarks. Noteworthy is, the order among the FTL algorithms is almost the same with various m when considering $\mathbf{E}[APY]$, but vary very little with $\mathbf{MED}[APY]$. In addition, RAPS and UCR are always in the midfield of the algorithms. SCR and UBH are showing always the worst APY. Beating UBH in terms of APY means "beating the market". It can be concluded, that all considered online algorithms, except SCR, are able to beat the market. But, only the FTL algorithms are able to beat powerfully the market with rising m. ONS requires also a large m to have a better APY.

Figure 6.3 illustrates the risk of algorithms measured by $\mathbf{E}[ASTDV]$ and $\mathbf{E}[MDD]$ on 1000 random market instances with various m. The investment horizon is again $T = 7564$. The results are ordered by $\mathbf{E}[ASTDV]$ and starting with the smallest value. Since the risk should be reduced smaller values of $\mathbf{E}[ASTDV]$ and $\mathbf{E}[MDD]$ are better than higher. First, the FTL algorithms are having the most risk. Their extent of risk increases with enlarging m. But, the risk of the other algorithms decreases in this case. It can be concluded, the FTW algorithms, risk algorithms and benchmarks benefit from diversification. In contrast, when considering some sample portfolio allocations of the FTL algorithms, these tend in the vast number of observations to a switching strategy where only one single asset is in the portfolio. This explains why they are so risky. CRAPS is the ALG with the lowest risk. However, CRAPS is always closely followed by RAPS, UCR, EG, UP and UBH. SCR and ONS are always in the midfield of algorithms for

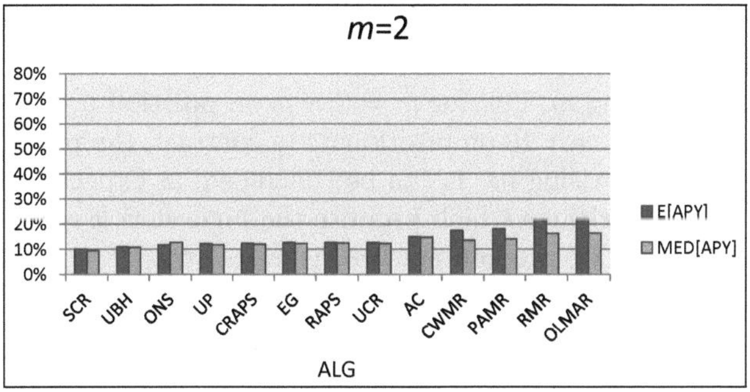

(a)

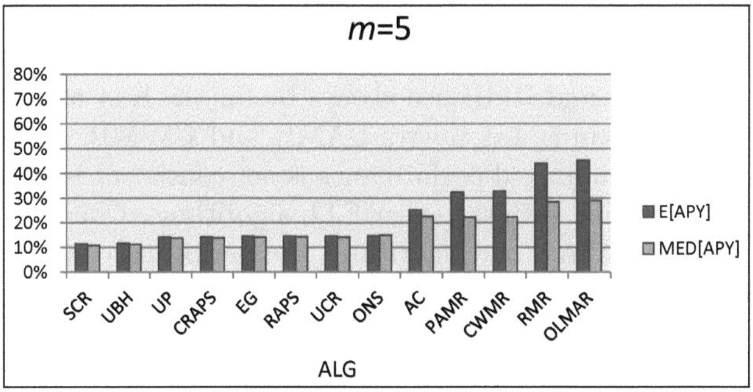

(b)

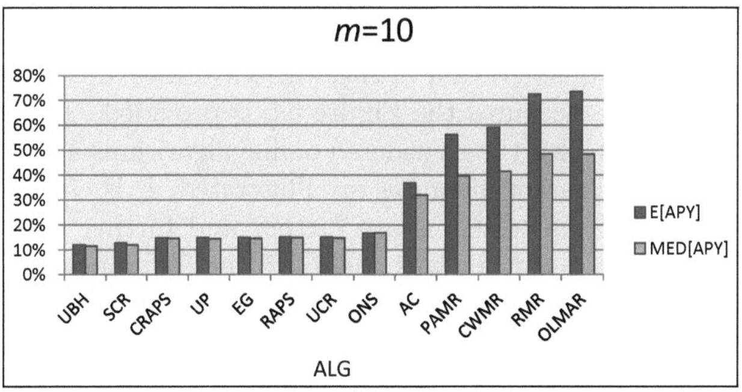

(c)

Figure 6.2: $\mathbf{E}[APY]$ (dark gray) and $\mathbf{MED}[APY]$ (light gray) for the considered online algorithms and benchmark algorithms for 1000 random markets with: (a) $m = 2$ assets, (b) $m = 5$ assets and (c) $m = 10$ assets

various m. In general, when $\mathbf{E}[ASTDV]$ increases then $\mathbf{E}[MDD]$ increases as well. Remarkable is, the benchmark UCR improves among the other algorithms with rising m, i.e., for $m = 2$ it is on position 6, for $m = 5$ it is on position 5 and for $m = 10$ on position 3. In contrast, the position of UBH is getting worse with rising m. It can be concluded, as expected, RAPS and CRAPS are the algorithms which promise the highest risk reduction.

The results with 1000 random market instances (with $T = 7564$) to evaluate the risk-adjusted performance of the considered algorithms is provided in Figure 6.4. The results are in an increasing order by $\frac{\mathbf{E}[APY]}{\mathbf{E}[ASTDV]}$. In addition, the results of $\frac{\mathbf{E}[APY]}{\mathbf{E}[MDD]}$ are given and would lead to a marginal different ranking. SCR is the worst ALG for all considered m, regardless on the specified statistical measure. The performance of ONS is also disappointing. Furthermore, SCR and ONS are clearly dominated by the other algorithms. In contrast, OLMAR and RMR are always by far the best algorithms. The next best two algorithms are always PAMR and CWMR. The only FTL algorithm whose risk-adjusted performance is not consistent is AC. But, with increasing m it is in line with the other FTL algorithms. Consider the benchmarks, in comparison to other algorithms the performance of UBH is poor and the performance of UCR is mainly in the midfield. The performances of RAPS and CRAPS are characterized that they are always between the FTW and FTL (except AC) algorithms. UP and EG are the best FTW algorithms and EG dominates always UP. Summing up, the results indicate that the FTL algorithms promise the largest risk-adjusted performance, followed by the risk algorithms.

Beside the single consideration of the statistical measures depending on specific performance measures the Figure 6.5 is provided. The $\mathbf{E}[APY]$ in relation to $\mathbf{E}[ASTDV]$ for the considered online algorithms and benchmarks for the 1000 random market instances are illustrated. It is clear to see, with increasing m the FTL algorithms are further and further away from the FTW algorithms. In addition, with rising m the differences of UCR, UP, EG, RAPS and CRAPS vanish. Another finding is very visible in this view, the FTW (FTL) algorithms have less (more) risk with increasing m.

As an intermediate result the following can be concluded: An investor who focuses on maximizing the return on investment or risk-adjusted performance should apply a FTL algorithm. More specifically, the OLMAR and the RMR are promising the best results. In contrast, an investor who focuses on the minimization of risk should choose RAPS or CRAPS.

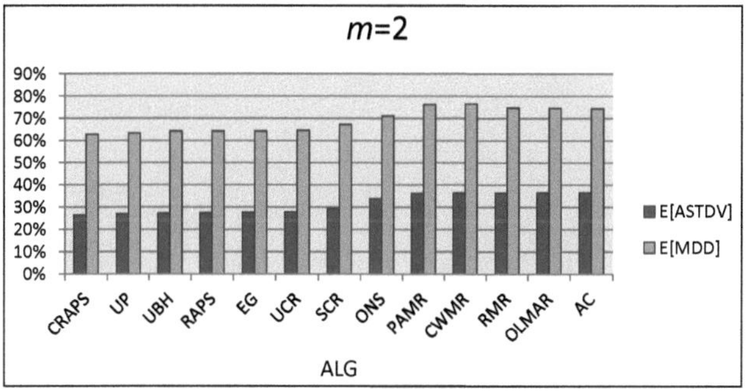

(a)

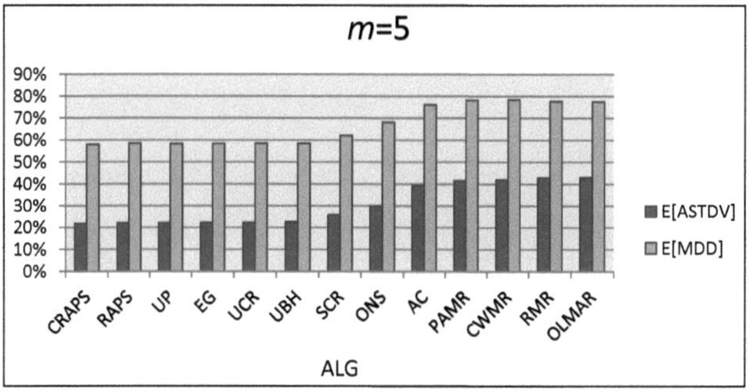

(b)

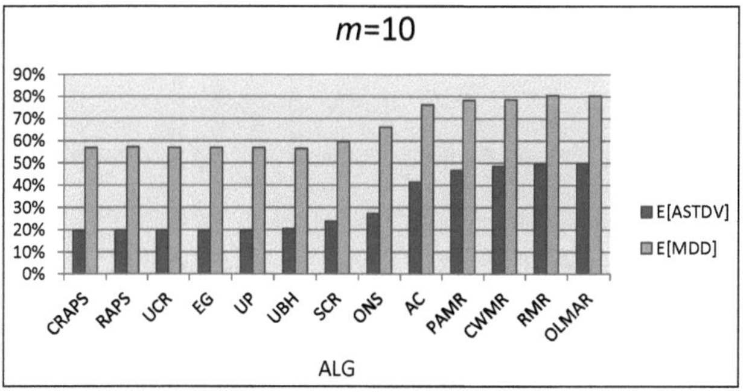

(c)

Figure 6.3: $\mathbf{E}[ASTDV]$ (dark gray) and $\mathbf{E}[MDD]$ (light gray) for the considered online algorithms and benchmark algorithms for 1000 random markets with: (a) $m = 2$ assets, (b) $m = 5$ assets and (c) $m = 10$ assets

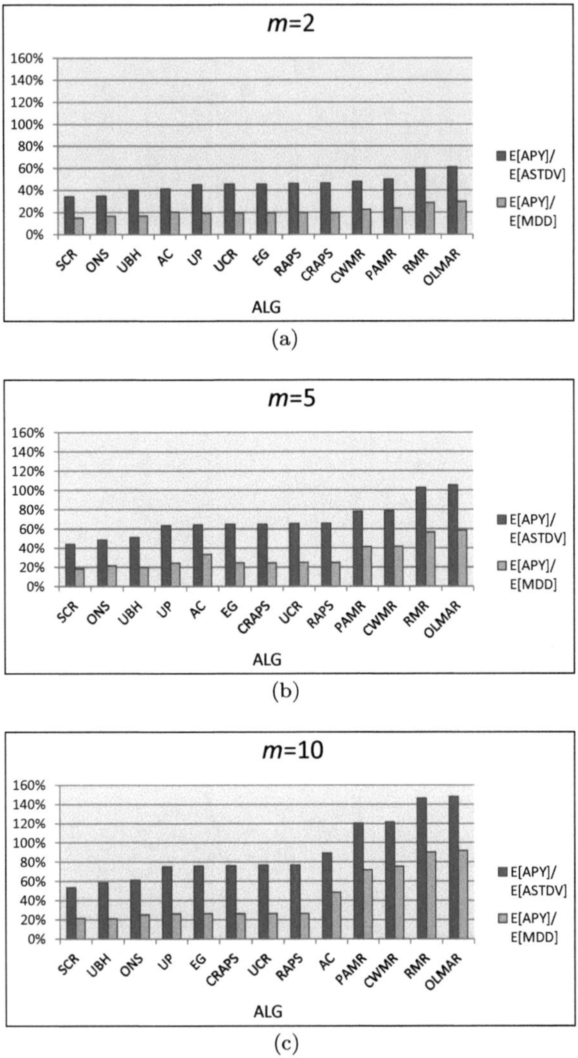

Figure 6.4: $\frac{\mathbf{E}[APY]}{\mathbf{E}[ASTDV]}$ (dark gray) and $\frac{\mathbf{E}[APY]}{\mathbf{E}[MDD]}$ (light gray) for the considered online algorithms and benchmark algorithms for 1000 random markets with: (a) $m = 2$ assets, (b) $m = 5$ assets and (c) $m = 10$ assets

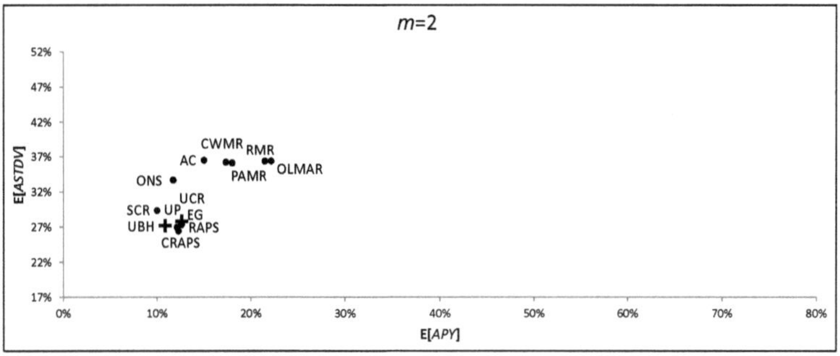

(a)

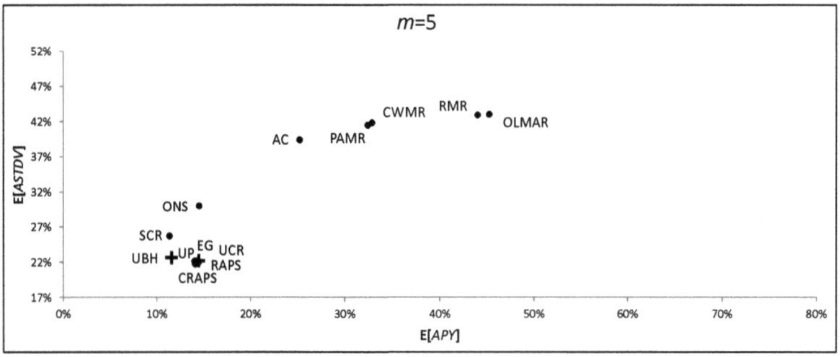

(b)

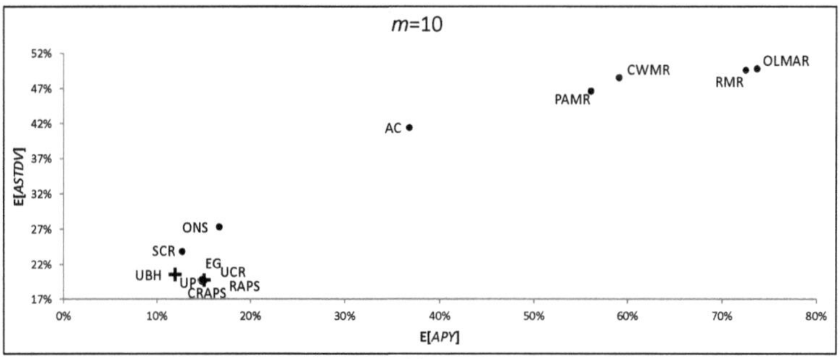

(c)

Figure 6.5: $\mathbf{E}[APY]$ in relation to $\mathbf{E}[ASTDV]$ for the considered online algorithms (spots) and benchmark algorithms (crosses) for 1000 random markets with: (a) $m = 2$ assets, (b) $m = 5$ assets and (c) $m = 10$ assets

6.4 Numerical Results: Beating the Benchmark

Recall, POO quantifies how often an ALG is able to beat the benchmark significantly on the considered sub-datasets. Figure 6.6 shows that the POO with performance measure APY for $m = 2$ and $m = 5$ assets in the market. As benchmarks the UBH and UCR are applied. The investment horizon is five years with $T = 1250$. When focusing on the benchmark UBH, the algorithms UP, EG, UCR, CRAPS and RAPS are able to beat UBH significantly with a $POO \approx 90\%$. In general, the FTW algorithms (except SCR) and the risk algorithms are having a higher potential to beat UBH than the FTL algorithms. In contrast, when the objective is to beat UCR then this finding alternates. The FTL algorithms are much better than most of the FTW algorithms and risk algorithms. Surprisingly, the ONS is the best ALG for $m = 2$ and the third best for $m = 5$ when the objective is to beat UCR. However, the POO of ONS reduces with rising m. The best FTL algorithms to beat both benchmarks are OLMAR and RMR. In addition, RAPS and CRAPS are almost equal to each other on a high level when the objective is to beat UBH, but CRAPS is absolutely poor when the objective is to beat UCR. Further, the POO against UBH for UP, EG and CRAPS is $POO \approx 90\%$ and for UCR it reduces to $POO < 10\%$. The reduction of the POO for RAPS is not that much, it is $POO = 22.22\%$.

Figure 6.7 shows the POO with the performance measure ASTDV. The conditions are the same as for the investigation for APY. As a first finding it can be stated, all FTL algorithms are never able to have a significantly lower ASTDV than UBH and UCR. On the contrary, RAPS and CRAPS are the best algorithms. Note, when the objective is to beat UCR then CRAPS and RAPS are able to generate a $POO \geq 90\%$. UP and EG are also able to generate a lower ASTDV than UBH and UCR, but for UCR on a substantially lower level than RAPS and CRAPS. It is easier to beat UCR in terms of ASTDV than UBH.

Figure 6.8 shows the POO with the performance measure RVR and the conditions are the same as for the investigations of the two previous performance measures. First, consider the objective to beat UBH. The algorithms UP, RAPS and CRAPS are able to generate significantly a larger RVR with a $POO > 90\%$. This is almost independent of m. It follows ONS which is very sensitive to m. After that, the FTL algorithms with OLMAR and RMR on top are following. The FTL algorithms promise a larger POO with rising m. The SCR is the worst ALG and is almost not

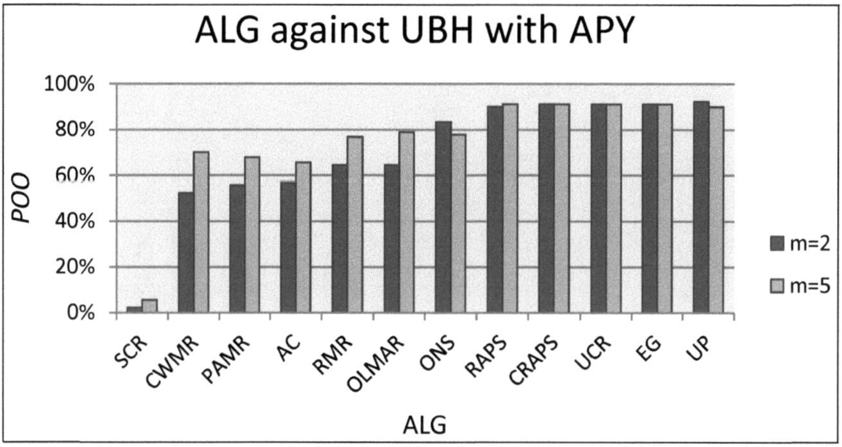

(a)

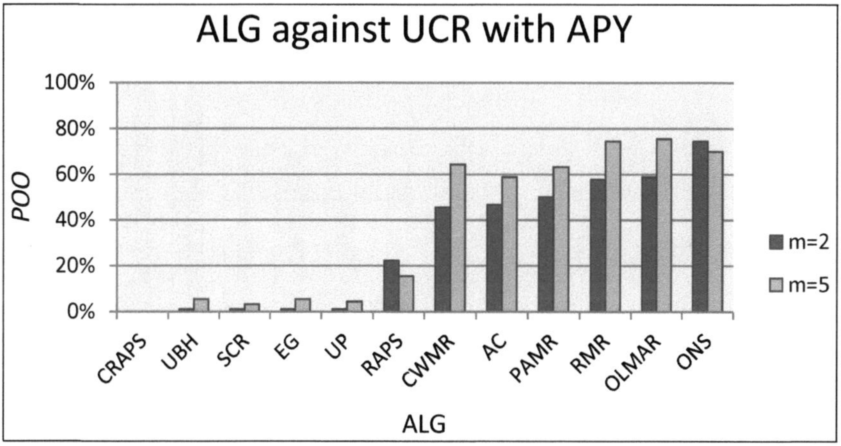

(b)

Figure 6.6: The illustration of *POO* with performance measure APY and investment horizon $T = 1250$ against benchmark (a) UBH and (b) UCR with $m = 2$ (dark gray) and $m = 5$ (light gray) assets in the market.

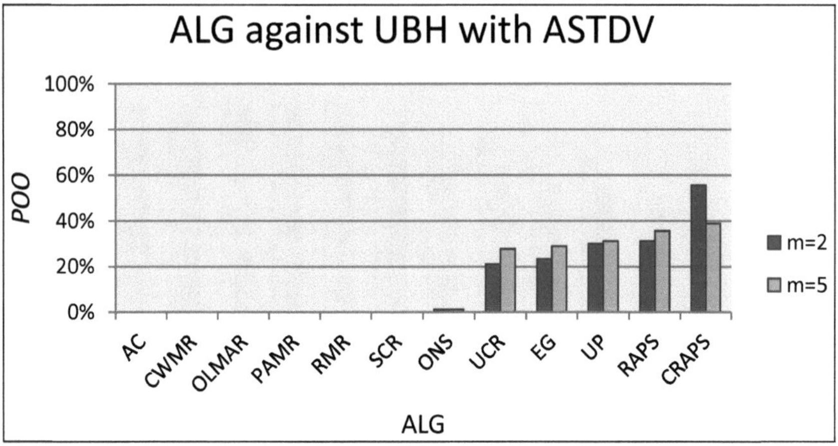

(a)

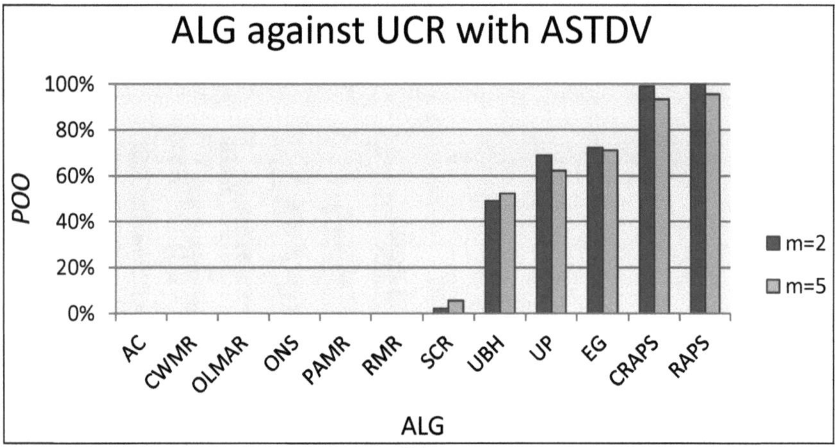

(b)

Figure 6.7: The illustration of *POO* with performance measure ASTDV and investment horizon $T = 1250$ against benchmark (a) UBH and (b) UCR with $m = 2$ (dark gray) and $m = 5$ (light gray) assets in the market.

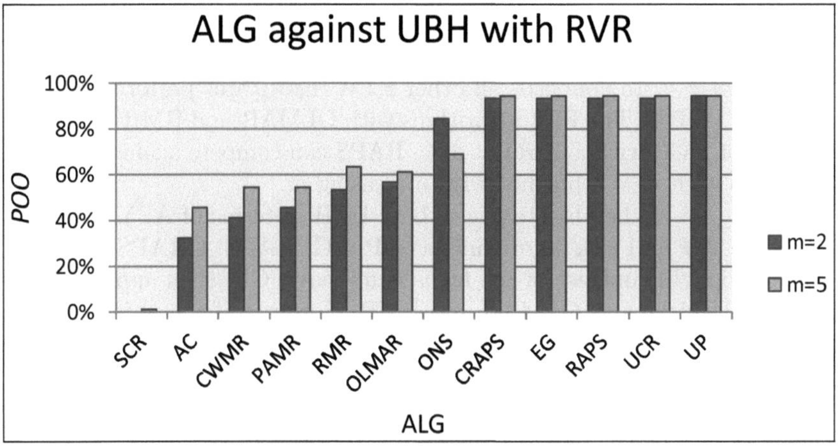

(a)

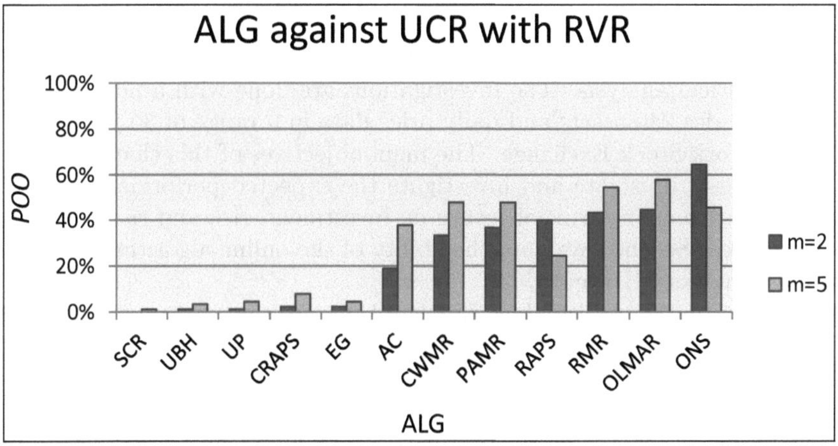

(b)

Figure 6.8: The illustration of *POO* with performance measure RVR and investment horizon $T = 1250$ against benchmark (a) UBH and (b) UCR with $m = 2$ (dark gray) and $m = 5$ (light gray) assets in the market.

able to beat UBH in terms of RVR. Next, consider the objective to beat UCR. Surprisingly, ONS with $m = 2$ is the best algorithm, but is only the fifth with $m = 5$. In contrast, all other FTW algorithms perform very low with $POO < 10\%$. The FTL algorithms with OLMAR and RMR on top are able to beat UCR with a $POO \approx 40\%$. RAPS can compete against the FTL algorithms with $m = 2$ but not with increasing m.

Summing up, if the objective is to beat UBH in terms of APY and RVR, then the FTW and risk algorithms as UP, EG, RAPS, CRAPS and ONS are beneficial. In contrast, if the focus is to beat UCR, then apply ONS or the FTL algorithms as OLMAR and RMR. However, if the objective is to beat UBH and UCR in terms of ASTDV, then always employ RAPS and CRAPS.

6.5 Conclusions

This chapter focused on the empirical performance of selected online portfolio selection algorithms in comparison to the benchmarks UBH and UCR based on a statistical analysis. The investigations are done with a new dataset, which includes 240 assets and daily price data in a range of 30 years from the New York Stock Exchange. The main objectives of this chapter are as follows: First, illustrate and investigate the expected performance of the online algorithms in terms of return on investment, risk and risk-adjusted performance. Second, evaluate the ability of the online algorithms to beat the benchmarks UBH and UCR.

Summing up the results, the FTL algorithms are the most promising algorithms for the maximization of return on investment and the maximization of risk-adjusted performance. In addition, the FTL algorithms are also able to beat UCR in these terms. More specifically, the OLMAR and RMR are the most encouraging online algorithms. The domination of FTL algorithms increases when the number of assets in the market goes up. In this respect, the previous results from the literature can be confirmed. In contrast, if the objective is to minimize the risk, then the risk algorithms RAPS and CRAPS are beneficial, especially when the objective is to have a lower risk than the UCR. The results for the FTW algorithms are disappointing. Although, they are able to beat UBH for all considered performance measures, but they are clearly dominated by the FTL algorithms and risk algorithms. An online ALG with a theoretical performance guarantee in the worst-case is apparently not beneficial to real markets.

The following comments concerning the interpretation of the executed investigations can be made and are taken into account for future work:

- **Parameter Setting Bias**: In previous investigations there is little evidence for critical parameter estimation for parametric online algorithms. In this work, the values for the parameters are taken directly from the literature and are independent of the new introduced dataset.
- **Asset and Time Period Selection Bias**: Following [Li et al., 2012, pp. 252,253], there is evidence that test designs are created to support the performance of a concrete ALG and that datasets are often reused. This bias has been addressed by two different approaches. (i) Collecting a larger dataset and producing sub-datasets as suggested by [Lo and MacKinlay, 1990, pp. 461-465] and [Hsu and Kuan, 2005, p. 608]. (ii) Choosing assets and starting points of the investment horizon randomly as in line with [Alpaydin, 2010, chap. 19].
- **Market Assumption Bias**: Transaction costs and taxes are not included in this work. It can be claimed, adding such assumptions worsen the performance of all online algorithms. FTL algorithms tend to switching portfolios, but keep an asset for longer periods in the portfolio such that not every trading period requires a transaction. In contrast, FTW and risk management strategies usually triggers several transactions in each trading period. Whether transactions costs and taxes would change the ranking between the online algorithms is an open question.
- **Survival Bias**: Following [Li et al., 2012, pp. 252,253], historical investigations suffer the drawback of choosing only assets that survived in hindsight, but crashing assets have mainly vanished. This problem could not be eliminated.
- **Market Liquidity Bias**: Following Li and Hoi [2014, p. 29], the executed empirical investigations claim that each asset can be quickly bought or sold. This may sometimes be not the case in practice and can only be tackled completely if the algorithms are applied in real trading.
- **Market Efficiency**: The applied online algorithms in the executed investigations take only historical information based on asset prices into account. The extraordinary "good" risk-adjusted performance of the FTL algorithms is striking. At first glance the results could be related to the weak EMH (see also Section 3.1.2). In line with Pearce [1987, p. 18], a test for the weak EMH requires an adequate and correct market model

that explains the price development of assets. However, especially with an online perspective on the PSP one tries to generate decisions without such a market model. The extent to which the results of the FTL algorithms can contribute to the discussion on the weak EMH must be done with more specific test designs as for example described in [Pearce, 1987, pp. 20-23].

7 A Software Tool for Testing Algorithms

A software tool to evaluate the empirical performance of an algorithm for the portfolio selection problem is introduced in this chapter. The complete software tool and all of its features are developed within the writing process of this thesis, in order to facilitate the execution of experiments to evaluate algorithms with various parameters on different datasets. It is a prototype of a trading system for backtesting, which should encourage researchers and practitioners to develop further software tools.

7.1 Preliminaries

In line with Kersch and Schmidt [2011], the evaluation of an online ALG for the PSP can be supported by a trading system which includes the automated execution of backtesting. The aim is to assess the practical applicability of an ALG on real markets. Intuitively, from a simplified perspective an ALG is practical when a large return on investment and a low variation in the return on investment (i.e., a large risk-adjusted performance) can be expected. Since markets are unpredictable such expectation and variation should be estimated on a variety of datasets. Such estimates should always be taken with caution because historical findings are not necessarily true for the future. However, they give an impression of what can be expected from an ALG when the future acts similarly to the past.

This chapter introduces the software tool *portfolio selection tool* (PST), which is designed to execute a large number of experiments with different algorithms and datasets. PST provides a graphical user interface and allows a user-friendly testing of the implemented algorithms. It runs on the operating system Microsoft Windows (7, 8 and 8.1) and is coded in Visual Basic.NET (framework 4.5). Note, the PST is far away from being complete and there is a lot of potential for improvements. Nevertheless, the software tool was designed to facilitate a fast empirical verification of findings during the creation of this thesis. Until now, very few literature is available with an empirical focus on online portfolio selection. More precisely, literature with comprehensive statistical analysis, test designs and datasets which are independent of a concrete ALG are missing.

Datasets with historical asset prices are widely provided as text files or in the form of databases. They are editable by the support of spreadsheet programs, e.g., Microsoft Excel. However, the creation of new sub-datasets by cutting an initial dataset into several parts can be a frustrating and time-consuming task. This operation is done to generate various market instances to test an ALG on different scenarios. Fortunately, this type of operation is automated in PST. A quick testing of an ALG and an illustration of results with different parameters can facilitate the understanding of an ALG for researchers and practitioners. Hence, the testing of one or more algorithms on an arbitrary number of given datasets is greatly simplified by clicking a button. In addition, an automated statistical analysis for the performance evaluation of the algorithms is included. To allow a further statical analysis it is possible to copy numerical results and paste them into an arbitrary spreadsheet program.

There is very few scientific literature on the design of a suitable trading system for portfolio selection as a software tool. However, the few that have been found are taken into consideration during the development of the PST. Kersch and Schmidt [2011] separate trading systems into (i) execution systems, (ii) planning systems and (iii) combined systems. In addition, a comparison of prominent trading systems from practice and a list of features for a suitable trading system are provided. Mitra et al. [2003] model the flow of information in a trading system which is built especially for portfolio selection. They call this trading system " integrated decision support system". It includes an automated data analysis and an integrated solver architecture to determine an "optimal" portfolio for an investor. Izumi et al. [2009] provide a framework to test algorithms for portfolio selection on artificial market instances which can be easily implemented in a trading system. Dong et al. [2004] introduce a web-based portfolio selection system for individual investors and a general architecture for arbitrary trading systems. Tucnik [2010] states that in general a trading system is basically only the implementation of an ALG in a software tool and lists the pros and cons of currently working trading systems in practice. Overall, it would be useful for researchers and practitioners if there is more scientific work on the design of suitable trading systems as a comprehensive software tool.

It follows a description of the primary functions which are included in PST.

7.2 Primary Functions

Before starting the PST the "Regions and Settings" of the operating system must be set to "English (US)" and an initial dataset is required. Such dataset is in the following called *universe*. It is a text file which consists of historical or artificial asset prices. It can be easily generated by a spreadsheet program or text editor. The text file must be in the structure of a comma-separated value file (the file name extension is often ".csv") and should be placed in an own folder on the hard drive. The structure of the universe file is as follows. The first line provides the header with "PriceNumber, Date, Asset1, Asset2, ...". The asset names can be chosen arbitrarily but without any comma within the name. Each following line consists of the price number which is equivalent to t (with $t = 0, 1, 2, ..., T$). The effective date of the taken asset prices must be provided in the form "month/day/year". The decimal sign in the asset prices has to be a dot (".").

When starting PST a graphical user interface as a window with six components appears. It consists of a (i) title bar, (ii) menu bar, (iii) scenario group container, (iv) scenario container, (v) scenario details container and (vi) status bar. In Figure 7.1 a screenshot of the PST when it is started is given. The title bar allows to change the size and position of the window and to close the PST. The menu bar contains all buttons and text fields to control the functions of the PST. The scenario group container is a placeholder for groups of scenarios. A scenario consists always exclusively of one market instance and a scenario group of one or more market instances. Selected scenarios for further calculations of an ALG or for the performance evaluation are displayed in the scenario container. When one scenario is selected from the scenario container by a doubleclick, a graphical illustration is provided in the scenario details container. The status bar shows information concerning the initial dataset and the current status when an ALG is calculated.

In the following, the primary functions of the PST are presented.

7.2.1 Executing Sampling

For the first time after a universe is loaded at least one market instance must be generated as input for the calculation and performance evaluation of an ALG. This is done by one of the implemented sampling techniques, by clicking one of the buttons in the sampling menu (see Figure 7.2). Each one generates a new folder into the folder where the universe is placed as a container for the new generated market instance. One market instance represents one scenario with at least $m \geq 2$ assets and $T \geq 2$ trading

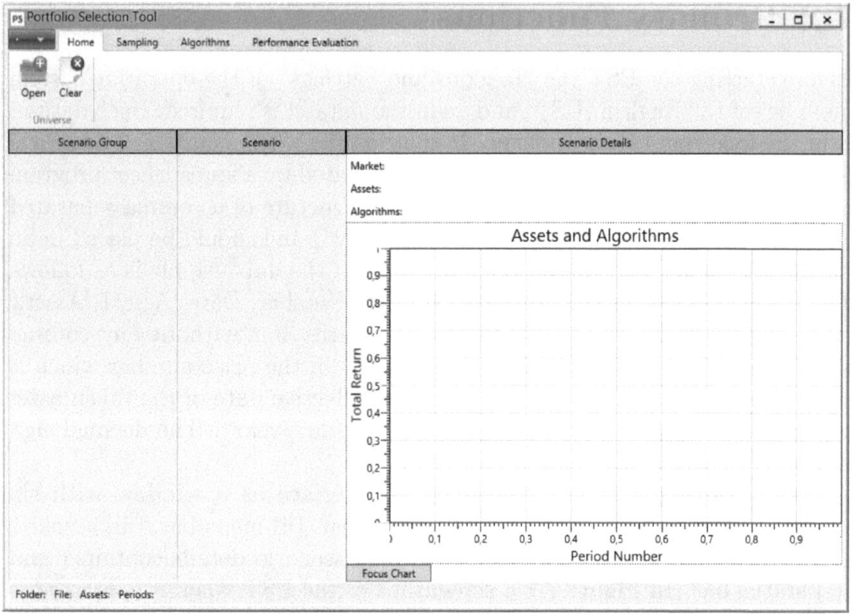

Figure 7.1: Main window of the software tool PST after starting

periods. The following sampling techniques to generate market instances are implemented:

(i) **All Data**: The entire universe is taken as one market instance.

(ii) **Full Enumeration**: Given is the number of desired assets in one market instance which must be lower or equal than the number of assets in the universe. For each possible asset combination one market instance is generated. Each market instance contains as many price numbers as available in the universe. Note, the number of combinations can be very large if there are a lot assets in the universe.

(iii) **Sampling with Assets**: In order to bypass the problem that the number of combinations can be very large when executing the full enumeration sampling technique, it is possible to generate only a given number of samples. For this, combinations are taken randomly. In the text field "# Samples" the number of required samples and in the text field "# Assets" the number of assets in one market instance can be manipulated.

(iv) **Sampling with Assets and Prices**: To investigate various investment horizons and to incorporate different starting dates of an ALG an integer length for the investment horizon must be given. It should be much lower than the available number of prices for each asset in the universe. Then, the randomization of (iii) plus a random starting price date generates one random market instance. In total, a given number of market instances are generated by this operation. Use the text field "# Prices" to set the number of prices which should be included in one market instance.

(v) **Sampling with Assets, Prices and Universe**: Given are two integer values ($cols$, $rows$) to separate the universe into $cols \times rows$ sub-universes, i.e., sub-datasets. The variable $cols$ separates the number of assets of the universe into $cols$ groups. The variable $rows$ cut the investment horizon of the universe into $rows$ non-overlapping time intervals. The remaining assets (price numbers) which are not a multiple of $cols$ ($rows$) are ignored. Finally, all sub-universes are having an equal number of assets and prices. On each sub-universe, operation (iv) is automatically executed. The text fields "# Columns" and "# Rows" are employed to set $cols$ and $rows$.

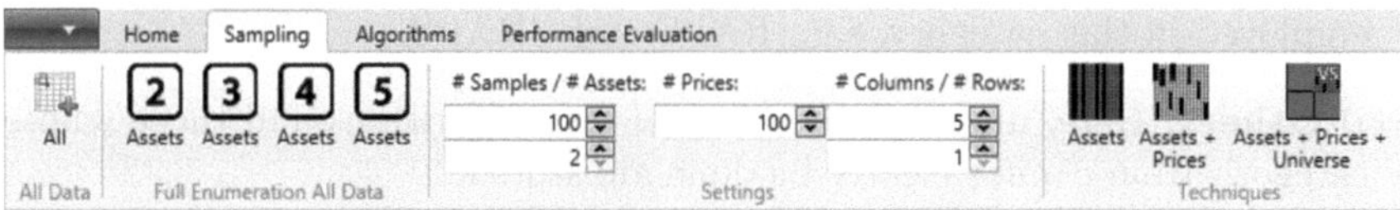

Figure 7.2: Menu "Sampling" of the software tool PST

It is possible to create a market instance manually. Then, generate a text file with the name "prices.csv" and manually build the market instance in the same structure as the file for the universe. Open the folder with the universe file and create a new folder which is a container for all manually created market instances. Put each manually created market instance as an own folder in this container. When the universe has been reloaded, PST recognizes the new market instance(s).

7.2.2 Running Algorithms

All algorithms for the PSP which are mentioned in this thesis are implemented in PST. There is a distinct button for each ALG. If the button is selected, then the concrete ALG is calculated automatically for all selected market instances which are listed in the scenario container. As seen in Figure 7.3 the algorithms are categorized by:

(i) **Non-Learning Benchmarks**: This category contains algorithms which are not influenced by the market itself. Thus, the decision for the portfolio allocation(s) is independent of the concrete market sequence, e.g., UBH, UCR and RAND.

(ii) **Offline Benchmarks**: Here, all algorithms are stored which require future information, i.e., offline algorithms. A practical application of these algorithms by a real investor is not possible. In general, these are divided into BH algorithms and CR algorithms. Some examples are BA, BCR and BVBH.

(iii) **Follow-the-Winner Algorithms**: The algorithms described by Li and Hoi [2014] where the universality can be proven are placed here, e.g., SCR, UP and EG.

(iv) **Follow-the-Loser Algorithms**: The algorithms described by Li and Hoi [2014] which exploit mean reversion are in this category, e.g., AC, PAMR and OLMAR.

(v) **Risk Algorithms**: Algorithms which incorporate risk management are employed in this category, e.g., RAPS and CRAPS.

(vi) **Others**: Here, uncompleted ideas of new algorithms and those whose categorization cannot clearly be done are stored.

(vii) **Packages**: If more than one ALG should be calculated by one click, then this operation is deposited here. Assume that one wants to calculated more than one ALG on the selected market instances, e.g., all algorithms from the Follow-the-Winner category.

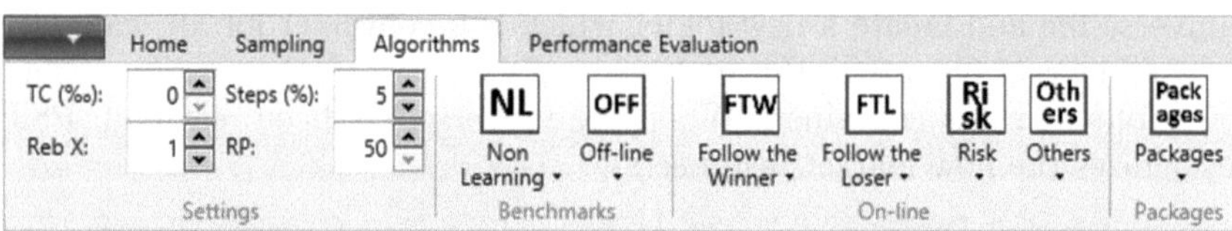

Figure 7.3: Menu "Algorithms" of the software tool PST

Till now it is not possible to add a new ALG without editing the source code of the PST via Visual Basic.NET. When executing an ALG the problem setting can be manipulated by adding transaction costs in the field "TC (‰)". At each moment where rebalancing is required the difference between the portfolio allocation before and after rebalancing is the basis for calculating the transaction costs. In addition, when manipulating the value of the text field "Reb X" algorithms with periodical rebalancing rebalance only each second, third, etc. trading period. The two additional text fields "Steps (%)" and "RP" control the number of experts for the calculation of experts algorithms. If the concrete market instance consists of two assets ($m = 2$), then the value of the text field "Steps (%)" represents the distance between the portfolio allocations of two consecutive non-learning experts (see Section 3.3). Problematic is, with increasing m the number of possible portfolio allocations and therefore the required computing time for an experts ALG increases exponentially (see Lemma A.3). To solve this problem, random experts with constant rebalancing using random portfolio allocations (see also Section 3.3.3) are applied for market instances with more than two assets ($m > 2$). The value of the text field "RP" represents the number of random experts. Some of the algorithms require additional parameters for the calculation. In this case, the values are requested in extra windows before the calculation is started.

7.2.3 Measuring Performance

When one or more algorithms are calculated on at least one market instance, the performance of the algorithms should be reported, analyzed and evaluated automatically. As it can be seen in Figure 7.4 in the menu bar of PST the following operations for the performance evaluation are available:

(i) **Market Details**: The details for the current selected market instance is shown. More specifically, the prices of the market instance, the portfolio allocations and the period wealth of all calculated algorithms are provided in a tabular form.

(ii) **Market Comparison**: From all selected market instances information of the consisting assets is extracted. Each asset of each market instance is considered as a single asset BH strategy. Its terminal wealth, exponential growth rate and standard deviation is calculated. In addition, to investigate the impact of correlation between return factors on the performance of an ALG the correlation coefficient of the return factors of the first two assets ($i = 1, 2$) of each concrete market instance is provided.

(iii) **Algorithm Comparison**: Information of all calculated algorithms of all market instances in one scenario group is extracted. All performance measures of Section 3.2 and statistical measures as arithmetic mean and standard deviation (see also Section 3.4.1) are included. Beyond that, one arbitrary already calculated ALG can be chosen as a benchmark. Based on the selected markets, it can be shown via hypothesis testing whether an ALG is significantly "better" than the benchmark in terms of a desired performance measure (see also Section 3.4.2).

(iv) **Universe Comparison**: An initial dataset is separated into multiple sub-datasets. On each dataset random market instances are generated (the sampling technique "Sampling with Assets, Prices and Universe" can be executed). At least two different algorithms must be calculated on all random market instances. The performance of the one ALG as test algorithm is tested against the other ALG as benchmark on each of the sub-datasets via hypothesis testing based on one desired performance measure (see also the second test design in Section 6.2).

Recall Figure 7.4, PST is shown with the selected menu "Performance Evaluation" and the following data can be seen: The loaded universe is placed in the folder "myData" and the corresponding text file is called "universe.csv". It includes 36 assets and 5652 trading periods. The folder "myData" includes one scenario group with the name "Popular Markets" and is also selected. It includes the four market instances "coke - ibm", "comme - kinar", "comme - meico" and "iroqu - kinar". For the market instance "comme - meico" detailed information is provided. The total returns of the underlying assets ("comme" and "meico") plus that of the non-learning algorithm UCR are illustrated in a graph.

7.3 Conclusions

In this chapter, a software tool, named *portfolio selection tool* (PST), is introduced. As a proof of concept for a trading system as a planning system it allows a quick testing of algorithms for the PSP on an arbitrary dataset with asset prices. This software tool facilitates the understanding of and the access to the algorithms of the literature for researchers and practitioners. The effort to generate new market instances to evaluate the performance of algorithms is significantly reduced. The primary functions of the software tool are described and illustrated by screenshots. The practicability of PST can be confirmed by the author through continuous use and further development as a part of this thesis.

During the practical application of PST the need of further functions arises, which should be addressed by the future development of this software

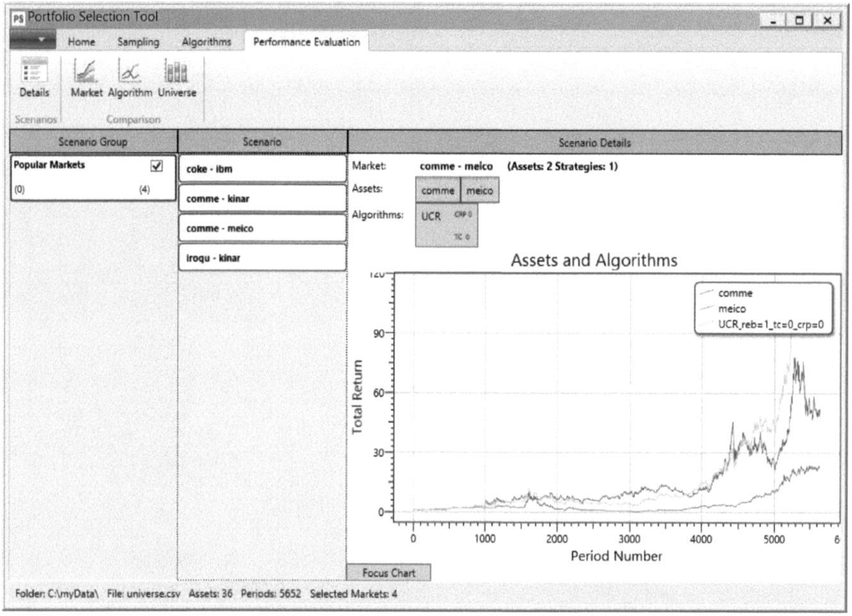

Figure 7.4: PST with Menu "Performance Evaluation" when one scenario group with four market instances is selected ("coke - ibm", "comme - kinar", "comme - meico" and "iroqu - kinar") and one market instance is selected for details illustration ("comme - meico")

tool: It is beneficial to have more graphical illustration of the data and the numerical results during the performance evaluation. This enhances the understanding of algorithms and reduces the dependence on spreadsheet programs. In addition, it would be nice to have a function that allows the implementation of new algorithms in the PST without manipulating the source code.

8 Conclusions and Future Work

The main scientific findings of this thesis are summarized in this chapter. The three research questions, which have been introduced in the first chapter are conclusively discussed and open issues for future work are given. In short, the exclusive scientific findings of this thesis are as follows:

(i) A simplified classification structure of selected types of the portfolio selection problem is mathematically derived.

(ii) Two competitive online algorithms with risk management, but exponential time complexity, are provided and analytically evaluated (risk-adjusted portfolio selection algorithm and combined risk-adjusted portfolio selection algorithm).

(iii) Prominent online algorithms of the literature from the machine learning community and the two new algorithms are empirically evaluated by a comprehensive statistical analysis. As results, the follow-the-loser algorithms are the most promising algorithms when the objective is the maximization of return on investment and risk-adjusted performance. The two new algorithms with risk management show excellent performance when the objective is the minimization of risk.

(iv) A prototype of a software tool for the automated statistical analysis of an online algorithm is created and described.

8.1 Portfolio Selection Problems

This thesis started with an introduction of relevant terms concerning the *portfolio selection problem* (PSP)[9] based on considerations of the *finance community* (FC) and the *machine learning community* (MLC). The online and offline perspective on the PSP is taken into account. Selected types of the PSP are classified and their relationships to each other are shown. More

[9]In this chapter abbreviations are reintroduced, in order to facilitate the understanding of a reader who is exclusively interested in the results of this thesis.

precisely, a *general portfolio selection problem* (GPSP) is formulated and the *constant rebalancing problem* (CRP), *buy-and-hold problem* (BHP) and *conversion problem* (CP), for example, are subsumed under it (Chapter 2).

In a further step, important performance measures and benchmarks are presented to quantify the concrete performance of an *algorithm* (ALG) for an arbitrary type of PSP. Motivated by the literature, they are the basis for the performance evaluation of an ALG by a statistical analysis or a competitive analysis. Within a complete chapter (Chapter 3), both concepts are described in detail. In addition, a prototype of a software tool for the automated statistical analysis of an online ALG is created and described (Chapter 7).

Summing up, Question 1 of Section 1.2 ("How to classify selected types of the PSP and evaluate the performance of an algorithm for the PSP?") is comprehensively answered. Nevertheless, there are some open issues for future work. In this thesis, short-selling is always excluded. It would be more realistic to model some types of the PSP with the ability to execute short-selling or at least with the possibility of buying some assets on credit. In addition, including consumption during the investment process would lead to additional types of the PSP. To model a PSP with inflows and outflows of wealth during the investment horizon is possible. Thus, the connections between the PSP and problems of financial planning could be shown. The use of online algorithms in financial planning is then only one step away.

Another open issue is the extension of the conversion problem to more than two assets. Consider the two cases: (i) At the beginning of the investment horizon there is only one base asset and the investor wants to allocate all wealth in the counter assets until the end of the investment horizon. In practice, this case occurs for example when an investor is starting with an investment process for the first time and all wealth is still invested in cash. Step by step the investor shifts wealth from the cash asset into the risky assets, i.e., conversion of the risk-free cash into the risky part of the market. At the end of the investment horizon, the investor is completely invested in the risky assets and a new PSP could be formulated without taking the cash asset into account. (ii) Consider the opposite case with only one counter asset. At the beginning of the investment horizon the wealth of the investor is entirely invested in the risky assets, i.e., the base assets. The aim is to allocate all wealth in the cash asset, as the only counter asset, until the end of the investment horizon. As an example, suppose that an investor retires and wants a complete elimination of risk in the portfolio. To solve the problems described by (i) and (ii) with competitive online algorithms, mathematical formulations of the problems are required.

8.2 Online Algorithms with Risk Management

In the further course of the thesis, prominent online algorithms for the PSP from the literature are presented with the support of pseudocodes (Chapter 4). The understanding of the underlying mechanisms of the provided algorithms is facilitated and should encourage researchers and practitioners to work more on them.

It turned out, risk management is little considered when online algorithms are designed in the MLC. This is addressed by Question 2 of Section 1.2 ("Is it possible to construct a worst-case competitive online algorithm for the PSP that takes risk management into account?"). Two online algorithms with risk management are proposed, i.e., *risk-adjusted portfolio selection algorithm* (RAPS) and *combined risk-adjusted portfolio selection algorithm* (CRAPS). Both have been proven by a competitive analysis that they are competitive in a Kelly market. This is a worst-case market where the prices of all assets decrease or remain constant during the entire investment horizon. In fact, the algorithms are modifications of the prominent *universal portfolio algorithm* (UP) and therefore based on experts. RAPS and CRAPS guarantee the same terminal wealth as UP for arbitrary instances of the Kelly market without making any statistical assumptions. As a result, Question 2 is answered positively (Chapter 5).

An open issue is the time complexity. Because experts are employed, the time complexity of RAPS and CRAPS is exponential and thus more work on online algorithms with risk management is required. To determine a "good" portfolio allocation, the concentration on experts is definitely not beneficial. A more promising type of construction for a faster online ALG with risk management could be motivated, for example, from the *exponential gradient algorithm* (EG) or *online Newton step algorithm* (ONS). With each trading period, the risk (risk-adjusted performance) of the portfolio converges to a historical optimal allocation only based on first and/or second order information (see Section 4.1.3).

8.3 Empirical Testing

Recall, there are many proposed online algorithms in the literature of the MLC, but there is only little comprehensive and independent statistical analysis of them. The focus on risk and risk-adjusted performance plus the ability of beating the *uniform buy-and-hold portfolio* (UBH) or *uniform constant rebalancing portfolio* (UCR) is minimal to none so far. Although, one reason to apply online algorithms is to achieve a better performance than UBH and UCR. For this reason, Question 3 of Section 1.2 ("How is the empirical performance of selected online algorithms for the PSP, in comparison to a benchmark algorithm, on a given dataset, when the performance is measured for the return on investment, risk and risk-adjusted performance?") is derived.

In this thesis, a comprehensive statistical analysis is executed based on a new dataset with 240 risky assets and 30-years price data from the New York Stock Exchange (Chapter 6). The empirical investigations are limited to the *follow-the-winner* (FTW) algorithms, the *follow-the-loser* (FTL) algorithms and the two proposed algorithms with risk-management. The following conclusions can be drawn based on the obtained numerical results: There is a lot of evidence that FTL algorithms are beneficial in terms of return on investment and risk-adjusted performance. In this regard, the results from the literature can be confirmed. The algorithms with risk management are in fact able to reduce the risk in the portfolio and are still capable to generate a high return on investment, but not as much as the FTL algorithms. In general, the performance of the FTW algorithms is poor for any considered performance measure and thus their use in practice is more than questionable.

There are some open issues concerning the empirical performance of the considered algorithms. A definite recommendation for investors on the practical application of the algorithms cannot be given yet. Further empirical tests with more realistic test designs including transaction costs and taxes are required. In addition, the findings should be verified on a second large dataset taken from another stock exchange.

Nevertheless, the already obtained numerical results are promising and encourage to develop more online algorithms. Based on the findings, an online ALG is desirable that on the one side exploits mean reversion (as the FTL algorithms) and on the other side takes risk management into account (as the two proposed risk algorithms). It would be interesting to see how the performance of such an ALG is in comparison to the state-of-the-art algorithms.

8.4 Concluding Remarks

Online algorithms for the PSP is still a relatively young subject in the MLC and received little to no attention in the FC. The design of solutions for the

PSP in the FC is mainly based on statistical models that attempt to explain the behavior of asset prices in the market. It is refreshing to see that the MLC developed such solutions without these models. A greater exchange of knowledge and experience between the two communities is now desirable and could lead to fruitful results. It is hoped that this thesis supports for this exchange.

A Proofs

A.1 Bounds on the Number of Allocations

Lemma A.1 *The number of allocations* Ω *is bounded by* $\left(\frac{m+\frac{1}{a}-1}{\frac{1}{a}}\right)^{\frac{1}{a}} \leq \Omega \leq m^{\frac{1}{a}}$ *when the minimum distance* a *between two consecutive allocations decreases. Let* $m, \frac{1}{a} \in \mathbb{N}^+$.

Proof A.1 *Let* $a' = \frac{1}{a}$.

$$\begin{aligned}
\Omega &= \binom{m+a'-1}{a'} \\
&= \frac{(m+a'-1)}{a'} \cdot \frac{(m+a'-1)-1}{a'-1} \cdots \frac{(m+a'-1)-(a'-1)}{1} \\
&\leq m^{a'}
\end{aligned} \tag{A.1}$$

$$\Omega \geq \left(\frac{m+a'-1}{a'}\right)^{a'}$$

□

Lemma A.2 *The number of allocations* Ω *is bounded by* $\left(\frac{m+\frac{1}{a}-1}{m-1}\right)^{(m-1)} \leq \Omega \leq (\frac{1}{a}+1)^{(m-1)}$ *when the number of elements* m *of each allocation increases. Let* $m, \frac{1}{a} \in \mathbb{N}^+$.

Proof A.2 *Let* $a' = \frac{1}{a}$.

$$\begin{aligned}
\Omega &= \binom{m+a'-1}{m-1} \\
&= \frac{(m+a'-1)}{m-1} \cdot \frac{(m+a'-1)-1}{m-1-1} \cdots \frac{(m+a'-1)-(m-1-1)}{1} \\
&\leq (a'+1)^{(m-1)}
\end{aligned} \tag{A.2}$$

$$\Omega \geq \left(\frac{m+a'-1}{m-1}\right)^{(m-1)}$$

□

A.2 Asymptotic Behavior of the Number of Allocations

Lemma A.3 *The asymptotic growth of the number of allocations Ω is exponential with increasing m and decreasing a. Let $m, \frac{1}{a} \in \mathbb{N}^+$.*

Proof A.3 *Let $a' = \frac{1}{a}$. It is sufficiently to show that the asymptotic growth of the lower bounds of Ω for m and a is exponential. Thus, show that $\left(\frac{m+a'-1}{a'}\right)^{a'} \in O(k^{a'})$ and $\left(\frac{m+a'-1}{m-1}\right)^{(m-1)} \in O(k^m)$ with $k > 1$ when using the lower bounds of Ω as derived in Lemma A.1 and Lemma A.2.*

$$\begin{aligned} \left(\frac{m+a'-1}{a'}\right)^{a'} &\in O(k^{a'}) \\ &\le ck^{a'} \\ \frac{\left(\frac{m+a'-1}{a'}\right)^{a'}}{k^{a'}} &\le c \end{aligned} \tag{A.3}$$

If and only if $k \ge \frac{m+a'-1}{a'}$, then the last inequality is always true for all $a' = m, ..., \infty$ with $c = 1$.

$$\begin{aligned} \left(\frac{m+a'-1}{m-1}\right)^{(m-1)} &\in O(k^m) \\ &\le ck^m \\ \frac{\left(\frac{m+a'-1}{m-1}\right)^{(m-1)}}{k^m} &\le c \end{aligned} \tag{A.4}$$

If and only if $k \ge \frac{m+a'-1}{m-1}$, then the last inequality is always true for all $m = 1, ..., \infty$ with $c = 1$.
It is shown that for increasing m and increasing a' there exists always a constant c.

□

B Numerical Results

B.1 Numerical Results: Expected Performance

ALG	$\mathbf{E}[APY]$	$\mathbf{MED}[APY]$	$\mathbf{E}[ASTDV]$	$\mathbf{E}[MDD]$	$\frac{\mathbf{E}[APY]}{\mathbf{E}[ASTDV]}$	$\frac{\mathbf{E}[APY]}{\mathbf{E}[MDD]}$
UBH	10.86%	10.74%	27.18%	64.26%	39.94%	16.89%
UCR	12.62%	12.30%	27.81%	64.55%	45.39%	19.56%
SCR	9.99%	9.44%	29.33%	67.23%	34.07%	14.86%
UP	12.12%	11.77%	26.95%	63.37%	44.97%	19.13%
EG	12.55%	12.22%	27.59%	64.27%	45.50%	19.53%
ONS	11.71%	12.68%	33.69%	71.01%	34.75%	16.49%
AC	15.01%	14.82%	36.48%	74.38%	41.16%	20.18%
PAMR	18.00%	14.03%	36.09%	76.21%	49.88%	23.62%
CWMR	17.33%	13.53%	36.20%	76.51%	47.86%	22.65%
OLMAR	22.16%	16.34%	36.37%	74.61%	60.91%	29.69%
RMR	21.50%	16.18%	36.35%	74.74%	59.13%	28.76%
RAPS	12.62%	12.45%	27.36%	64.27%	46.11%	19.63%
CRAPS	12.27%	11.96%	26.42%	62.80%	46.45%	19.54%

Table B.1: Summary of numerical results for the expected performance with $m = 2$

ALG	**E**[APY]	**MED**[APY]	**E**[$ASTDV$]	**E**[MDD]	$\frac{\mathbf{E}[APY]}{\mathbf{E}[ASTDV]}$	$\frac{\mathbf{E}[APY]}{\mathbf{E}[MDD]}$
UBH	11.56%	11.17%	22.10%	58.29%	51.14%	19.77%
UCR	14.51%	14.91%	22.61%	58.50%	64.74%	24.57%
SCR	11.35%	10.74%	25.71%	62.02%	44.13%	18.29%
UP	14.10%	13.84%	22.02%	58.44%	63.96%	33.14%
EG	14.46%	14.28%	22.20%	58.45%	64.90%	24.35%
ONS	14.03%	13.69%	30.01%	68.03%	48.35%	21.33%
AC	25.19%	22.39%	43.01%	77.47%	63.47%	24.07%
PAMR	32.85%	22.24%	39.39%	76.01%	78.53%	41.92%
CWMR	32.41%	22.12%	41.47%	78.24%	78.15%	41.43%
OLMAR	45.25%	29.05%	42.91%	77.72%	105.23%	58.42%
RMR	44.02%	28.47%	41.83%	78.37%	102.58%	56.63%
RAPS	14.48%	14.14%	22.18%	58.42%	65.24%	24.77%
CRAPS	14.36%	14.02%	21.72%	57.89%	65.67%	24.75%

Table B.2: Summary of numerical results for the expected performance with $m = 5$

ALG	**E**[APY]	**MED**[APY]	**E**[$ASTDV$]	**E**[MDD]	$\frac{\mathbf{E}[APY]}{\mathbf{E}[ASTDV]}$	$\frac{\mathbf{E}[APY]}{\mathbf{E}[MDD]}$
UBH	12.72%	11.96%	19.72%	56.95%	60.99%	25.18%
UCR	16.66%	16.85%	20.51%	56.51%	75.90%	26.11%
SCR	11.95%	11.48%	23.84%	59.63%	53.35%	21.33%
UP	14.86%	14.50%	19.65%	57.13%	75.73%	26.30%
EG	15.05%	14.88%	19.78%	56.91%	76.42%	26.46%
ONS	14.83%	14.59%	27.32%	66.16%	58.26%	21.15%
AC	36.77%	31.91%	49.82%	80.36%	75.10%	26.11%
PAMR	59.10%	41.49%	41.39%	76.23%	121.73%	75.28%
CWMR	56.10%	39.69%	46.63%	78.27%	120.31%	71.67%
OLMAR	73.72%	48.38%	49.62%	80.58%	147.98%	91.73%
RMR	72.54%	48.39%	48.55%	78.50%	146.18%	90.02%
RAPS	15.07%	14.78%	19.77%	56.93%	76.60%	26.35%
CRAPS	14.97%	14.63%	19.54%	56.79%	88.83%	48.23%

Table B.3: Summary of numerical results for the expected performance with $m = 10$

B.2 Numerical Results: Beating the Benchmark

	$m = 2$			$m = 5$		
ALG	*APY*	*ASTDV*	*RVR*	*APY*	*ASTDV*	*RVR*
UCR	91.11%	21.11%	93.33%	91.11%	27.78%	94.44%
SCR	2.22%	0.00%	0.00%	5.56%	0.00%	1.11%
UP	92.22%	30.00%	94.44%	90.00%	31.11%	94.44%
EG	91.11%	23.33%	93.33%	91.11%	28.89%	94.44%
ONS	83.33%	1.11%	84.44%	77.78%	1.11%	68.89%
AC	56.67%	0.00%	32.22%	65.56%	0.00%	45.56%
PAMR	55.56%	0.00%	45.56%	67.78%	0.00%	54.44%
CWMR	52.22%	0.00%	41.11%	70.00%	0.00%	54.44%
OLMAR	64.44%	0.00%	56.67%	78.89%	0.00%	61.11%
RMR	64.44%	0.00%	53.33%	76.67%	0.00%	63.33%
RAPS	90.00%	31.11%	93.33%	91.11%	35.56%	94.44%
CRAPS	91.11%	55.56%	93.33%	91.11%	38.89%	94.44%

Table B.4: Summary of numerical results (POO) for beating UBH with $m = 2$ and $m = 5$

	$m = 2$			$m = 5$		
ALG	*APY*	*ASTDV*	*RVR*	*APY*	*ASTDV*	*RVR*
UBH	1.11%	48.89%	1.11%	5.56%	52.22%	3.33%
SCR	1.11%	2.22%	0.00%	3.33%	5.56%	1.11%
UP	1.11%	68.89%	1.11%	4.44%	62.22%	4.44%
EG	1.11%	72.22%	2.22%	5.56%	71.11%	4.44%
ONS	74.44%	0.00%	64.44%	70.00%	0.00%	45.56%
AC	46.67%	0.00%	18.89%	58.89%	0.00%	37.78%
PAMR	50.00%	0.00%	36.67%	63.33%	0.00%	47.78%
CWMR	45.56%	0.00%	33.33%	64.44%	0.00%	47.78%
OLMAR	58.89%	0.00%	44.44%	75.56%	0.00%	57.78%
RMR	57.78%	0.00%	43.33%	74.44%	0.00%	54.44%
RAPS	22.22%	100.00%	40.00%	15.56%	95.56%	24.44%
CRAPS	0.00%	98.89%	2.22%	0.00%	93.33%	7.78%

Table B.5: Summary of numerical results (POO) for beating UCR with $m = 2$ and $m = 5$

Bibliography

Agarwal, A., Hazan, E., Kale, S. and Schapire, R. E.: 2006, Algorithms for portfolio management based on the newton method, *Proceedings of International Conference on Machine Learning*, Vol. 23, pp. 9–16.

Ahmad, I., Mohr, E. and Schmidt, G.: 2014, Risk-adjusted on-line portfolio selection, *Operations Research Proceedings 2012*, Springer, pp. 63–69.

al-Binali, S.: 1999, A risk-reward framework for the competitive analysis of financial games, *Algorithmica* **25**(1), 99–115.

Albers, S.: 2006, Online algorithms, *Interactive Computation: The New Paradigm*, Springer, pp. 143–164.

Alpaydin, E.: 2010, *Introduction to Machine Learning*, 2 edn, MIT Press.

Arlot, S. and Celisse, A.: 2010, A survey of cross-validation procedures for model selection, *Statistics Surveys*, Vol. 4, pp. 40–79.

Bell, R. M. and Cover, T. M.: 1980, Competitive optimality of logarithmic investment, *Mathematics of Operations Research* **5**(2), 161–166.

Bertsekas, D. P.: 1982, Projected newton methods for optimization problems with simple constraints, *SIAM Journal on Control and Optimization* **20**(2), 221–246.

Black, F.: 1971, Implications of the random walk hypothesis for portfolio management, *Financial Analysts Journal* **27**(2), 16–22.

Blum, A.: 1998, On-line algorithms in machine learning, *Online Algorithms: The State of the Art*, Springer, chapter 14, pp. 306–325.

Blum, A. and Kalai, A.: 1999, Universal portfolios with and without transaction costs, *Machine Learning* **35**(3), 193–205.

Blum, W. and Niss, M.: 1991, Applied mathematical problem solving, modelling, applications, and links to other subjects - state, trends and issues in mathematics instruction, *Educational Studies in Mathematics* **22**(1), 37–68.

Borodin, A. and El-Yaniv, R.: 1998, *Online computation and competitive analysis*, Vol. 2, Cambridge University Press.

Borodin, A., El-Yaniv, R. and Gogan, V.: 2000, On the competitive theory and practice of portfolio selection (extended abstract), *LATIN 2000: Theoretical Informatics* **1776**, 173–196.

Borodin, A., El-Yaniv, R. and Gogan, V.: 2004, Can we learn to beat the best stock, *Journal of Artifical Intelligence Research* **21**, 579–594.

Brock, W., Lakonishok, J. and LeBaron, B.: 1992, Simple technical trading rules and the stochastic properties of stock returns, *Journal of Finance* **47**(5), 1731–1764.

Burns, P.: 2006, Random portfolios for performance measurement, *Optimisation, Econometric and Financial Analysis*, Vol. 9, E. Kontoghiorghes and C. Gatu, pp. 227–249.

Campbell, R., Huisman, R. and Koedijk, K.: 2001, Optimal portfolio selection in a value-at-risk framework, *Journal of Banking & Finance* **25**(9), 1789–1804.

Cesa-Bianchi, N. and Lugosi, G.: 2006, *Prediction, Learning, and Games*, Cambridge University Press.

Chen, N.-F., Roll, R. and Ross, S. A.: 1986, Economic forces and the stock market, *Journal of Business* **59**(3), 383–403.

Chen, Y. and Ye, X.: 2011, Projection onto a simplex. Working Paper.

Clarke, R., de Silva, H. and Thorley, S.: 2006, Minimum-variance portfolios in the U.S. equity market, *Journal of Portfolio Management* **33**(1), 10–24.

Cohen, K. J. and Pogue, J. A.: 1967, An empirical evaluation of alternative portfolio-selection models, *Journal of Business* **40**(2), 166–193.

Conrad, J., Dittmar, R. F. and Ghysels, E.: 2013, Ex ante skewness and expected stock returns, *Journal of Finance* **68**(1), 85–124.

Cont, R.: 2001, Empirical properties of asset returns: stylized facts and statistical issues, *Quantitative Finance* **1**(2), 223–236.

Copeland, T. E.: 1976, A model of asset trading under the assumption of sequential information arrival, *Journal of Finance* **31**(4), 1149–1168.

Cover, T. M.: 1991, Universal portfolios, *Mathematical Finance* **1**(1), 1–29.

Cover, T. M. and Gluss, D. H.: 1986, Empirical bayes stock market portfolios, *Advances in Applied Mathematics* **7**(2), 170–181.

Cover, T. M. and Ordentlich, E.: 1996, Universal portfolios with side information, *IEEE Transactions on Information Theory* **42**(2), 348–363.

Crammer, K., Dekel, O., Keshet, J., Shalev-Shwartz, S. and Singer, Y.: 2006, Online passive-aggressive algorithms, *Journal of Machine Learning Research* **7**(3), 551–585.

Crammer, K., Dredze, M. and Pereira, F.: 2008, Exact convex confidence-weighted learning, *Advances in Neural Information Processing Systems 21*, pp. 345–352.

Dannoura, E. and Sakurai, K.: 1998, An improvement on El-Yaniv-Fiat-Karp-Turpin's money-making bi-directional trading strategy, *Information Processing Letters* **66**(1), 27–33.

Dantzig, G. B.: 1948, Programming in a linear structure, *Comptroller*, Vol. 17, USAF Washington D.C.

Darvas, Z.: 2009, Leveraged carry trade portfolios, *Journal of Banking & Finance* **33**(5), 944–957.

DeBondt, W. F. and Thaler, R.: 1985, Does the stock market overreact?, *Journal of Finance* **40**(3), 793–805.

DeMiguel, V., Garlappi, L. and Uppal, R.: 2009, Optimal versus naive diversification: How inefficient is the 1/n portfolio strategy?, *Review of Financial Studies* **22**(5), 1915–1953.

DeMiguel, V., Utrera, A. M. and Nogales, F. J.: 2015, Parameter uncertainty in multiperiod portfolio optimization with transaction costs, *Journal of Financial and Quantitative Analysis* p. in press.

Dochow, R., Mohr, E. and Schmidt, G.: 2014, Risk-adjusted on-line portfolio selection, *Operations Research Proceedings 2013*, Springer, pp. 113–119.

Dong, J., Du, H. S., Wang, S., Chen, K. and Deng, X.: 2004, A framework of web-based decision support systems for portfolio selection with OLAP and PVM, *Decision Support Systems* **37**(3), 367–376.

Dredze, M., Crammer, K. and Pereira, F.: 2008, Confidence-weighted linear classification, *Proceedings of the International Conference on Machine Learning*, Vol. 25, pp. 264–271.

Efron, B.: 1979, Bootstrap methods: Another look at the jackknife, *Annals of Statistics* **7**(1), 1–26.

Efron, B. and Gong, G.: 1983, A leisurely look at the bootstrap, the jackknife, and cross-validation, *The American Statistican* **37**(1), 36–48.

El-Yaniv, R.: 1998, Competitive solutions for online financial problems, *ACM Computing Surveys* **30**(1), 28–69.

El-Yaniv, R., Fiat, A., Karp, R. M. and Turpin, G.: 2001, Optimal search and one-way trading online algorithms, *Algorithmica* **30**(1), 101–139.

Elton, E. J. and Gruber, M. J.: 1997, Modern portfolio theory, 1950 to date, *Journal of Banking & Finance* **21**(11), 1743–1759.

Estrada, J.: 2006, Downside risk in practice, *Journal of Applied Corporate Finance* **18**(1), 117–125.

Fama, E. F.: 1970a, Efficient capital markets: A review of theory and empirical work, *Journal of Finance* **25**(2), 383–417.

Fama, E. F.: 1970b, Multiperiod consumption-investment decisions, *American Economic Review* **1**(1), 163–174.

Fama, E. F. and French, K. R.: 1992, The cross-section of expected stock returns, *Journal of Finance* **47**(2), 427–465.

Fourer, R., Gay, D. M. and Kerninghan, W. B.: 1993, *AMPL: A Modeling Language for Mathematical Programming*, 2 edn, Scientic Press.

Fujiwara, H., Iwama, K. and Sekiguchi, Y.: 2011, Average-case competitive analyses for one-way trading, *Journal of Combinatorial Optimization* **21**(1), 83–107.

Gaivoronski, A. A. and Stella, F.: 2000, Stochastic nonstationary optimization for finding universal portfolios, *Annals of Operations Research* **100**(1), 165–188.

Garlappi, L., Uppal, R. and Wang, T.: 2007, Portfolio selection with parameter and model uncertainty: A multi-prior approach, *Review of Financial Studies* **20**(1), 41–81.

Geltner, D.: 1997, The use of appraisals in portfolio valuation and index construction, *Journal of Property Valuation & Investment* **15**(5), 423–447.

Gibbons, M. R., Ross, S. A. and Shanken, J.: 1989, A test of the efficiency of a given portfolio, *Econometrica* **57**(5), 1121–1152.

Godino, J. D. and Batanero, C.: 1998, Clarifying the meaning of mathematical objects as a priority area for research in mathematics education, *Mathematics Education as a Research Domain: A Search for Identity*, Vol. 4 of *New ICMI Studies Series*, Springer, pp. 177–195.

Györfi, L., Ottucsk, G. and Urban, A.: 2012, Empirical log-optimal portfolio selections: a survey, *Machine Learning for Financial Engineering*, World Scientific, chapter 2, pp. 79–115.

Györfi, L., Urban, A. and Vajda, I.: 2007, Kernel-based semi-log-optimal empirical portfolio selection stratgies, *International Journal of Theoretical and Applied Finance* **10**(3), 505–516.

Hakansson, N. H.: 1970, Optimal investment and consumption strategies under risk for a class of utility functions, *Econometrica* **38**(5), 587–607.

Hakansson, N. H.: 1974, Convergence to isoelastic utility and policy in multiperiod portfolio choice, *Journal of Financial Economics* **1**(3), 201–224.

Helmbold, D., Schapire, R., Singer, Y. and Warmuth, M.: 1998, On-line portfolio selection using multiplicative updates, *Mathematical Finance* **8**(4), 325–347.

Hoe, L. W., Hafizah, J. S. and Zaidi, I.: 2010, An empirical comparison of different risk measures in portfolio optimization, *Business and Economic Horizons* **1**(1), 39–45.

Horst, R. and Tuy, H.: 1996, *Global Optimization: Deterministic Approaches*, 3 edn, Springer.

Hsu, P.-H. and Kuan, C.-M.: 2005, Reexamining the profitability of technical analysis with data snooping checks, *Journal of Financial Econometrics* **3**(4), 606–628.

Huang, D., Zhou, J., Li, B., Hoi, S. C. and Zhou, S.: 2013, Robust median reversion strategy for on-line portfolio selection, *Proceedings of the 23th International Joint Conference on Artificial Intelligence*, Vol. 23, pp. 2006–2012.

Ishijima, H.: 2001, Numerical methods for universal portfolios. Working Paper.

Izumi, K., Toriumi, F. and Matsui, H.: 2009, Evaluation of automated-trading strategies using an artificial market, *Neurocomputing* **72**(16), 3469–3476.

Jegadeesh, N. and Titman, S.: 1993, Returns to buying winners and selling losers: Implications for stock market efficiency, *Journal of Finance* **48**(1), 65–91.

Jensen, M. C.: 1969, Risk, the pricing of capital assets, and the evaluation of investment portfolios, *Journal of Business* **42**(2), 167–247.

Jones, C. M. and Lamont, O. A.: 2002, Short-sale constraints and stock returns, *Journal of Financial Economics* **66**(2), 207–239.

Kanji, G. K.: 2006, *100 Statistical Tests*, 3 edn, SAGE Publications.

Karatzas, I., Lehoczky, J. P. and Shreve, S. E.: 1987, Optimal portfolio and consumption decisions for a small investor on a finite horizon, *SIAM Journal on Control and Optimization* **25**(6), 1557–1586.

Karp, R. M.: 1992, On-line algorithms versus off-line algorithms: how much is it worth to know the future?, *Proceedings of the IFIP Twelfth World Computer Congress*, pp. 416–429.

Kelly, J. L.: 1956, A new interpretation of information rate, *Bell System Technical Journal* **35**(4), 917–926.

Kersch, M. and Schmidt, G.: 2011, Survey of trading systems for individual investors, *Information Systems for Global Financial Markets: Emerging Developments and Effects*, IGI Global, chapter 8, pp. 157–176.

Knuth, D. E.: 1968, *The Art of Computer Programming*, Vol. 1, Addison-Wesley.

Koutsoupias, E. and Papadimitriou, C. H.: 2000, Beyond competitive analysis, *SIAM Journal on Computing* **30**(1), 300–317.

Kozat, S. S. and Singer, A. C.: 2011, Universal semiconstant rebalanced portfolios, *Mathematical Finance* **21**(2), 293–311.

Kraus, A. and Litzenberger, R. H.: 1976, Skewness preference and the valuation of risk assets, *Journal of Finance* **4**(4), 1085–1100.

Lee, C. F.: 1977, Functional form, skewness effect, and the risk-return relationship, *Journal of Financial and Quantitative Analysis* **12**(1), 55–72.

Li, B. and Hoi, S. C. H.: 2012, On-line portfolio selection with moving average reversion, *Proceedings of the 29th International Conference on Machine Learning*, pp. 1–8.

Li, B. and Hoi, S. C. H.: 2014, Online portfolio selection: A survey, *ACM Computing Surveys* **46**(3), 1–35.

Li, B., Hoi, S. C. H. and Gopalkrishnan, V.: 2011, Corn: Correlation-driven nonparametric learning approach for portfolio selection, *ACM Transactions on Intelligent Systems and Technology* **2**(3), 1–29.

Li, B., Hoi, S. C. H., Zhao, P. and Gopalkrishnan, V.: 2013, Confidence weighted mean reversion strategy for online portfolio selection, *ACM Transactions on Knowledge Discovery from Data* **7**(1), 1–38.

Li, B., Zhao, P., Hoi, S. C. H. and Gopalkrishnan, V.: 2012, PAMR: Passive aggressive mean reversion strategy for portfolio selection, *Machine Learning* **87**(2), 221–258.

Li, D. and Ng, W.-L.: 2000, Optimal dynamic portfolio selection: Multiperiod mean-variance formulation, *Mathematical Finance* **10**(3), 387–406.

Liberti, L., Cafieri, S. and Tarissan, F.: 2009, Reformulations in mathematical programming: A computational approach, *Foundations of Computational Intelligence*, Vol. 3, Springer, pp. 153–234.

Lisi, F.: 2011, Dicing with the market: Randomized procedures for evaluation of mutual funds, *Quantitative Finance* **11**(2), 163–172.

Lo, A. W. and MacKinlay, A. C.: 1990, Data-snooping biases in tests of financial asset pricing models, *Review of Financial Studies* **3**(3), 431–467.

Magdon-Ismail, M., Atiya, A. F., Pratap, A. and Abu-Mostafa, Y. S.: 2004, On the maximum drawdown of a Brownian motion, *Journal of Applied Probability* **41**(1), 147–161.

Malin, M. and Bornholt, G.: 2013, Long-term return reversal: Evidence from international market indices, *Journal of International Financial Markets, Institutions & Money* **25**, 1–17.

Malkiel, B. G.: 1973, *A Random Walk Down Wall Street*, Norton.

Malkiel, B. G.: 2003, The efficient market hypothesis and its critics, *Journal of Economic Perspectives* **17**(1), 59–82.

Manasse, M., McGeoch, L. and Sleator, D.: 1988, Competitive algorithms for on-line problems, *Proceedings of Annual ACM Symposium on Theory of Computing*, Vol. 20, pp. 322–333.

Markowitz, H. M.: 1952, Portfolio selection, *Journal of Finance* **7**(1), 77–91.

Mayo, H.: 2010, *Investments: An Introduction*, 10 edn, Cengage Learning.

McMillan, M. G., Pinto, J. E., Pirie, W. and de Venter, G. V.: 2011, *Investments - Principles of Portfolio and Equity Analysis*, Wiley.

Merton, R. C. and Samuelson, P. A.: 1974, Fallacy of the log-normal approximation to optimal portfolio decision-making over many periods, *Journal of Financial Economics* **1**(1), 67–94.

Mitra, G., Kyriakis, T., Lucas, C. and Pirbhai., M.: 2003, A review of portfolio planning: models and systems, *Advances in Portfolio Construction and Implementation*, chapter 1, pp. 2–39.

Modigliani, F. and Modigliani, L.: 1997, Risk-adjusted performance, *Journal of Portfolio Management* **23**(2), 45–54.

Mohr, E., Ahmad, I. and Schmidt, G.: 2014, Online algorithms for conversion problems: A survey, *Surveys in Operations Research and Management Science* **19**(2), 87–104.

Mohr, E. and Schmidt, G.: 2013, How much is it worth to know the future in online conversion problems?, *Discrete Applied Mathematics* **161**(10), 1546–1555.

Mossin, J.: 1966, Equilibrium in a capital asset market, *Econometrica* **34**(4), 768–783.

Mossin, J.: 1968, Optimal multiperiod portfolio policies, *Journal of Business* **41**(2), 215–229.

Nawrocki, D. N.: 1991, Optimal algorithms and lower partial moment: ex post results, *Applied Economics* **23**(3), 465–470.

Nee, V. and Opper, S.: 2010, Political capital in a market economy, *Social Forces* **88**(5), 2105–2132.

Ordentlich, E. and Cover, T. M.: 1998, The cost of achieving the best portfolio in hindsight, *Mathematics of Operations Research* **23**(4), 960–982.

Ott, L. and Longnecker, M.: 2008, *An Introduction to Statistical Methods and Data Analysis*, 6 edn, Cengage Learning.

Pearce, D. K.: 1987, Challenges to the concept of stock market efficiency, *Economic Review* **72**(8), 16–33.

Prix, J., Loistl, O. and Huetl, M.: 2007, Algorithmic trading patterns in Xetra orders, *European Journal of Finance* **13**(8), 717–739.

Quenouille, M. H.: 1949, Problems in plane sampling, *Annals of Mathematical Statistics* **20**(3), 355–357.

Rachev, S. T., Hoechstoetter, M., Fabozzi, F. J. and Focardi, S. M.: 2010, *Probability and Statistics for Finance*, Wiley.

Rader, D. J.: 2010, *Deterministic Operations Research: Models and Methods in Linear Optimization*, Wiley.

Richards, A. J.: 1997, Winner-loser reversals in national stock market indices: Can they be explained?, *Journal of Finance* **52**(5), 2129–2144.

Rockafellar, T. R. and Uryasev, S.: 2000, Optimization of conditional value-at-risk, *Journal of Risk* **2**(3), 21–42.

Rosen, K.: 2011, *Discrete Mathematics and its Applications*, 7 edn, McGraw-Hill.

Roy, B.: 1971, Problems and methods with multiple objective functions, *Mathematical Programming* **1**(1), 239–266.

Ryan, B.: 2007, *Corporate Finance and Valuation*, Cengage Learning.

Sadoghi, M., Singh, M. L. H., Shum, W. and Jacobsen, H.-A.: 2010, Efficient event processing through reconfigurable hardware for algorithmic trading, *Proceedings of the VLDB Endowment*, Vol. 3, pp. 1525–1528.

Schmidt, G., Mohr, E. and Kersch, M.: 2010, Experimental analysis of an online trading algorithm, *Electronic Notes in Discrete Mathematics* **36**, 519–526.

Sharpe, W. F.: 1963, A simplified model for portfolio analysis, *Management Science* **9**(2), 277–293.

Sharpe, W. F.: 1964, Capital asset prices: A theory of market equilibrium under conditions of risk, *Journal of Finance* **19**(3), 425–442.

Sharpe, W. F.: 1966, Mutual fund performance, *Journal of Business* **39**(1), 119–138.

Singer, Y.: 1997, Switching portfolios, *International Journal of Neural Systems* **8**(4), 445–455.

Smith, B. M.: 2005, Modelling for constraint programming, *CP Summer School 2005* .

Smith, R. T.: 2001, Price volatility, welfare, and trading hours in asset markets, *Journal of Banking & Finance* **25**(3), 479 – 503.

Spielman, D. A. and Teng, S.-H.: 2004, Smoothed analysis of algorithms: Why the simplex algorithm usually takes polynomial time, *Journal of the ACM* **51**(3), 385–463.

Stickel, S. E.: 1995, The anatomy of the performance of buy and sell recommendations, *Financial Analysts Journal* **51**(5), 25–39.

Tobin, J.: 1958, Liquidity preference as behavior toward risk, *Review of Economic Studies* **25**(2), 65–86.

Treynor, J. L.: 1965, How to rate management of investment funds, *Hardvard Business Review* **43**(1), 63–75.

Tsang, E.: 1993, *Foundations of Constraint Satisfaction*, Academic Press.

Tsay, R. S.: 2010, *Analysis of Financial Time Series*, 3 edn, Wiley.

Tucnik, P.: 2010, Optimization of automated trading systems interaction with market environment, *Lecture Notes in Business Information Processing*, Vol. 64, Springer, pp. 55–61.

Tukey, J. W.: 1958, Bias and confidence in not quite large samples, *Annals of Mathematical Statistics* **29**, 614.

Vath, V. L., Mnif, M. and Pham, H.: 2007, A model of optimal portfolio selection under liquidity risk and price impact, *Finance and Stochastics* **11**(1), 51–90.

Vecer, J.: 2006, Maximum drawdown and directional trading, *Risk* **19**(12), 88–92.

Vovk, V. G. and Watkins, C.: 1998, Universal portfolio selection, *Proceedings of the Annual Conference on Learning Theory*, Vol. 11, pp. 12–23.

Zhang, W., Xu, Y., Zheng, F. and Dong, Y.: 2012, Optimal algorithms for online time series search and one-way trading with interrelated prices, *Journal of Combinatorial Optimization* **23**(2), 159–166.

MIX
Papier aus verantwortungsvollen Quellen
Paper from responsible sources
FSC® C105338

If you have any concerns about our products,
you can contact us on
ProductSafety@springernature.com

In case Publisher is established outside the EU,
the EU authorized representative is:
Springer Nature Customer Service Center GmbH
Europaplatz 3, 69115 Heidelberg, Germany

Printed by Libri Plureos GmbH
in Hamburg, Germany